CHRONOLOGY OF ANCIENT EGYPT

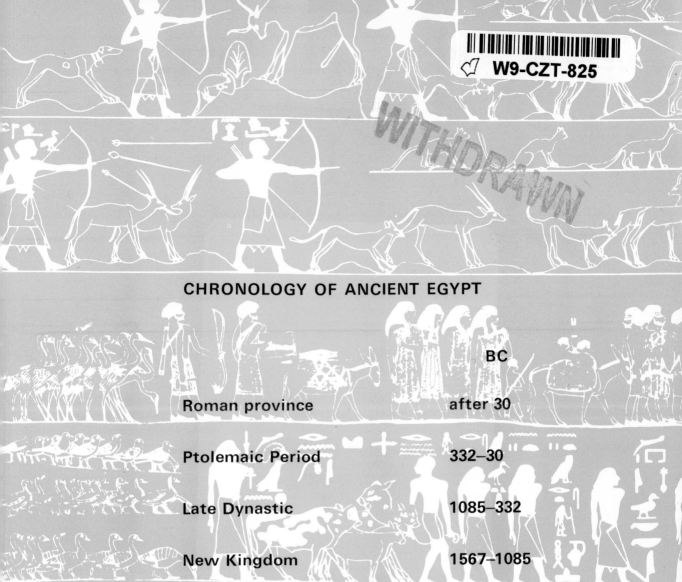

	BC
Roman province	after 30
Ptolemaic Period	332–30
Late Dynastic	1085–332
New Kingdom	1567–1085
Middle Kingdom	2181–1567
Old Kingdom	2686–2181
Early Dynastic Period	3100–2686
Pre-Dynastic Period	before 3100

Domesticated animals
from early times

Juliet Clutton-Brock

DOMESTICATED ANIMALS
from early times

University of Texas Press, Austin

British Museum (Natural History)

Published by the British Museum (Natural History), London and the University of Texas Press, Austin, Texas

© British Museum (Natural History) 1981

First University of Texas Press Edition, 1981

International Standard Book Number 0-292-71532-3

Library of Congress Catalog Card Number 80-71080

Printed in England
—

Contents

Acknowledgements

My first thanks are for Debbie Lyne of the Publications Section of the British Museum (Natural History) who assembled all the illustrations, and for Gordon Corbet and Chris Owen who read and edited the manuscript. I should also like to thank Richard Burleigh, Caroline Grigson, Colin Groves, Peter Jewell, and Ann and Gale Sieveking for their help and constructive comments.

Those people who kindly provided me with photographs are acknowledged in the text. I should like to thank them as well as Ray and Corinne Burrows for the line drawings of animals, Brian Groombridge who drew the skeletal material, Valerie Jones who drew the maps, and Gillian Greenwood who designed the text and jacket. The manuscript was typed by Joanna Bridges-Webb.

Preface

This book is an attempt to explore the manner in which man has manipulated and changed the way of life of other mammals since the end of the Ice Age, ten thousand years ago. Its main concern is with the enfoldment of animals within human societies, and with the archaeological and historical evidence for the early stages of domestication.

The common domestic mammals of the Old World were all well established as discrete, breeding populations isolated from their wild parent species by the time of the Roman Empire. I have therefore taken the arbitrary decision to conclude the description and natural histories of these domestic mammals with that period. Therefore there is no account here of the history of livestock improvement that began in Europe in the later Middle Ages. Neither is there discussion of pathology nor of the milk yields or carcass weights of modern commercial breeds, these details being readily available in publications on veterinary science and agriculture. On the other hand there are descriptions of those species of mammals that are still undergoing the process of domestication, the most successful of which may prove to be the elk and the eland. There is also a discussion of the exploitation of wild herbivores in the past and at the present day.

The views expressed are my own and if some appear to be contentious I can only hope that they will provide a stimulus for future work into the origins and social behaviour of domesticated animals.

NOTES 1. The conventional terms, Palaeolithic, Mesolithic, Neolithic, Bronze Age and Iron Age are used in this book in a general sense to describe the successive phases of hunting, primitive farming, and early metal-working communities. The terms should not, however, be seen as a chronological sequence because they were not necessarily synchronous or of the same duration in every region.

2. The present-day distributions of the less abundant wild mammals, quoted throughout the text, are summarized from the *Red Data Book* I Mammalia. IUCN, Morges, Switzerland, 1972. The terms, 'endangered', 'vulnerable', and 'rare' are used in accordance with the connotations of the *Red Data Book*, as are the zoogeographic regions of the world that are shown in the map (Fig. I.1) on the next page.

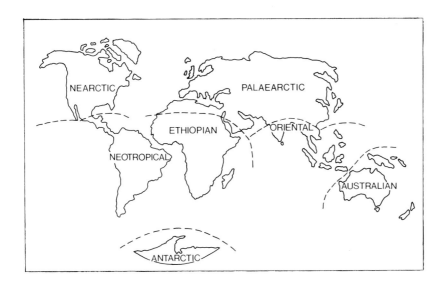

Figure I.1 Zoogeographical regions of the world

Introduction

Man's place in nature at the end of the Ice Age

Animals, especially mammals, have always been a part of human society. One reason for this being that, unlike the remaining living primates, man evolved as a carnivorous predator dependent on his mental and physical prowess to kill other animals for food. This entailed the development of complicated, social relationships between the hunters, their prey, and competing predators. The capacity for this kind of interaction remains with us today, for even the city-dweller has many behavioural characteristics and traits that reflect ancestral adaptation to the life of the hunting and gathering communities of the Ice Age, or Pleistocene period. Non-verbal communication, which is very highly developed in human beings, is perhaps the most outstanding of these traits. More readily than from his speech we can tell the mood of a neighbour by an often unconscious assessment of his facial expressions, posture, and bodily movements, and in the same way we can interpret the behaviour, attitudes, and feelings of many members of the animal world. It is this ability that has greatly helped man to enfold a wide variety of animal species within his own social organization.

At first, human hunters adapted their behaviour to that of their prey according to the biological processes that are inherent in all relationships between predator and prey. With some human societies this relationship evolved into cultural systems that became specialized for hunting selected species of large mammals, as happened with the American Indians who followed the herds of bison, or the Laplanders, some of whom are still dependent on reindeer for almost everything they need for survival. Other human groups, instead of bending their own activity towards that of their prey, learned how to manipulate or tame the behaviour of some of the animals that they could communicate with, and so began the process of domestication.

More than a hundred years ago, Francis Galton (1865) wrote a remarkable essay on the domestication of animals. The concluding paragraphs of this essay are quoted here because, although the wording would be different today, they still provide a comprehensive summary of what is known about the history of man's domination and manipulation of the animal kingdom:

I see no reason to suppose that the first domestication of any animal, except the elephant, implies a high civilisation among the people who established it. I cannot believe it to have been the result of a pre-conceived intention, followed by elaborate trials, to administer to the comfort of man. Neither can I think it arose from one successful effort made by an individual, who might thereby justly claim the title of benefactor to his race; but, on the contrary, that a vast number of half-unconscious attempts have been made throughout the course of ages, and that ultimately, by slow degrees, after many relapses, and continued selection, our several domestic breeds became firmly established.

I will briefly restate what appear to be the conditions under which wild animals may become domesticated; – 1, they should be hardy; 2, they should have an inborn liking for man; 3, they should be comfort-loving; 4, they should be found useful to the savages; 5, they should breed freely; 6, they should be easy to tend. [These 'conditions' are re-interpreted on pp. 15–16.]

It would appear that every wild animal has had its chance of being domesticated, that those few which fulfilled the above conditions were domesticated long ago, but that the large remainder, who fail sometimes in only one small particular, are destined to perpetual wildness so long as their race continues. As civilisation extends they are doomed to be gradually destroyed off the face of the earth as useless consumers of cultivated produce.

Each of Galton's six conditions expresses, admittedly in rather old-fashioned terms, the behavioural structure upon which man's association with other species of animal can be interpreted and they can therefore be discussed in relation not only to the history of domestication but also to the biology and behavioural patterns of man and animal.

Evolutionary change in wild animals or plants occurs as a result of natural selection in response to environmental, climatic, and other conditions combined with reproductive isolation. In the domestic animal or plant change occurs as a result of artificial selection by humans rather than by nature, and similarly reproductive isolation is maintained by the activities of man rather than by geographical barriers. However, evolution in any organism takes place as a result of biological processes and this applies to the human species as to any other living animal or plant, wild or domestic. It applies also to the evolution of human social and cultural systems (what Galton called civilisation) but what characterizes all aspects of human evolution and separates them from the non-human is the rapid rate of their development.

During the last 250 000 years (a short interval of time in evolutionary terms) the human species has increased in numbers from an estimated three million to about 4000 million and the life expectancy of the individual has increased over the same period from less than 30 years to over 70 years. The amazing increase in population was correlated with emigration away from the tropical regions where hominids first evolved and increasing colonization of the northern hemisphere during the Upper Pleistocene (Campbell, 1972). It did not take place as a gradual progression, as usually occurs under natural selection, but as a series of surges. The first of these happened when the emerging *Homo sapiens* learned how to make tools and manipulate fire which enabled man to survive in the periglacial conditions of northern Europe and America during the last glaciation. It was probably during this period that man first began to associate with the wolf, the progenitor of the dog. Perhaps they were even then hunting partners, for within recent years the excavation by H. de Lumley of a hillside cave (La Grotte du Lazaret) in the south of France has shown how Palaeolithic man*, about 125 000 years ago, built shelters within the cave, and a remarkable feature of these shelters was that each had the skull of a wolf intentionally placed at its entrance (Lumley, 1969)†. For the terminology of the Pleistocene and relative chronologies see the chart, Appendix II, p. 198.

The second great surge in human progress with its ensuing increase in population numbers occurred when man learned how to cultivate plants and tame and domesticate animals for his own use, whilst the third major surge took place with the intensive industrialization with which we are familiar today.

Each of these surges was a progression towards increasing mastery over the environment and pressures that control the evolutionary success or failure in all other forms of life. Despite man's ability to dominate the earth he is still, however, tied to physiological adaptations formed in his early hominid ancestors as a response to tropical climates and he is still tied to behavioural adaptations that evolved at a later stage in the hunters of large prey during the Ice Age. A unique and paradoxical feature of man is that he is a tropical, omnivorous primate whose exceptional success as a species began to accelerate only when he became a social hunter in a subarctic environment. Furthermore there is another paradox in that it was the behavioural structure of the social hunter and killer that enabled man to enfold other species of animal within his communities and to tame them and control their breeding to such an extent that many domesticated animals today bear little resemblance to their wild ancestors.

When discussing the associations that can be formed between human and animal communities it is necessary to understand what is really meant by the terms wildness, fierceness and tameness. The terms reflect relationships between man and animal that are relics of the attitudes and behaviour of man the hunter and master predator who, for perhaps the last half million years, has been the enemy of all other animals including the largest mammals, at first on the land but later in the sea as well. A *wild* animal is usually thought of as one that is either very fierce or very shy and runs away on sight. The more quickly it reacts, the wilder it is considered to be. So that animals in a deer park or nature reserve do not appear to be truly wild because they are used to the presence of humans. If the animals are not hunted they are not frightened of man and they do not run away, but

† L. R. Binford who has recently examined the animal remains from Lazaret claims that these skulls are from a wolf den in the cave, and are nothing to do with hominid ritual (unpublished, 1980).

* See note 1 on p. 7.

this does not mean that they are not wild, as long as they are living independently of human handling and control. Examples of this can be found in the descriptions of sailors and explorers who when arriving in newly discovered territory or on an uninhabited island have found the wild mammals and birds to show no fear whatsoever of the human invaders. As happened with the famous example of the Dodo in Mauritius the unfortunate animals are thought to be 'stupid' and have almost always suffered a speedy extermination.

Now that the majority of human communities are no longer primarily hunters for food and people only kill for 'sport' their attitudes to wild animals have changed and are complicated by an exaggerated belief in the *fierceness* of wild beasts. This follows from progressive estrangement and loss of contact with the animal world, combined with an emotional need to retain dominance over it. A *tame* animal differs from a wild one in that it is dependent on man and will stay close to him of its own free will. All mammals can be tamed if they are taken from their mothers and reared in association with a human protector from a very young age. Whether they will remain tame as adults depends to a great extent on the degree of development of their innate social behavioural patterns, that is whether they are solitary or social in their way of life.

The predatory mammal may be either a solitary hunter like the members of the cat family (with the exception of the lion) or the fox, partly social like the jackal or coyote, or highly social like the African hunting dog, wolf, or man. Which of these types of hunter the predator belongs to depends on the size of its prey – those carnivores that live on small prey, killing animals that are smaller than themselves, are solitary, whilst it is obvious that to attack an animal much larger than itself the predator has to use a team-effort. Also it would be wasteful if a carnivore such as a wolf could kill an animal the size of a bison every time it wanted a single meal, for this much meat would feed a whole group (Fig. I.2, p. 18).

Although man is a primate and the wolf belongs to the quite separate Order Carnivora, the life of the Palaeolithic hunters in the northern hemisphere during the Pleistocene period was much closer in style to that of the wolf than it was to any other member of the Order Primates. Both wolf and man evolved as social hunters and during the glacial phases of the Upper Pleistocene they had the same ubiquitous distribution and they preyed on the same herds of large mammals. Highly sophisticated social behaviour evolved in both man and wolf in response to the harsh conditions of life in the tundra where meat was plentiful nearly the whole year round but vegetable foods were in scant supply except in the peak of the arctic spring. But, whereas the true wolf, *Canis lupus*, had been an established member of the north European and Asian ecosystems for nearly a million years, man

Section I
Man-made animals

Figure I.2 Tundra wolf, *Canis lupus*, see p. 12 (photo Geoffrey Kinns).

Figure I.3 Mixed herds of livestock as they may be seen today in Iraq, see p. 20 (photo author).

Figure 1.1 Dogs in a camp of the Ritarrngo-speaking peoples of north-east Arnhem Land, Australia, to show the variation that may be found within one native population. In order to keep their food away from their dogs the people have to eat on specially constructed wooden platforms, see p. 22 (photo N. Peterson).

Man-made animals

Descriptions are given in this section of the species of mammal whose appearance and way of life has been most changed by their association with man. The process of domestication is outlined, and explanation is given of the terms used in the nomenclature and classification of domestic mammals. This is followed by the history of the domestication of the dog, an animal that holds a unique place in man's favour as ancient hunting partner and companion.

Approximately 9000 years ago the human hunter-gatherers in those parts of the world that were relatively densely populated, and notably in western Asia, began to alter their way of life. Instead of hunting wild animals for food they began to keep flocks

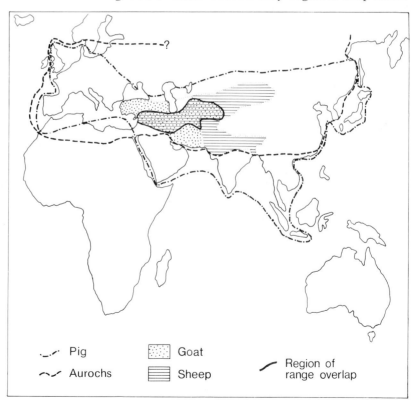

Figure I.4 Map to show the probable distributions of the progenitors of the four main species of domestic livestock, pig, aurochs, goat and sheep, at the end of the last Ice Age. It can be seen that all four were present in western Asia.

Pig
Aurochs
Goat
Sheep
Region of range overlap

and herds of livestock animals to kill when required, and to cultivate cereals and other plant foods (Fig. I.3, p. 18 and the map Fig. I.4).

Was this change a natural step forward in the development of human cultural systems or was it a response to the pressures for survival in a harsh environment where water was often scarce and the numbers of people were steadily increasing? Human beings probably require the challenge of increased population combined with demanding conditions in their environment before they will change their social structure. In reality, life in the 'Fertile Crescent' may have been very tough before the earliest farmers learned how to irrigate the land and manipulate the wild herbivores so that they provided a readily available walking-larder.

Sheep and goats were probably the first livestock animals to be domesticated, followed by cattle and pigs. The horse was the last of the group of man-made animals to be enfolded into human society, and yet economic progress has probably depended more on the horse than on any other animal. In the Old World the dispersal into new territory by invading peoples, until the invention of the motor engine, either peacefully or by force, was greatly aided by the horse and the mule.

1 *Domestication as a biological process*

A domestic animal is one that has been bred in captivity for purposes of economic profit to a human community that maintains complete mastery over its breeding, organization of territory, and food supply. Not all adult mammals can flourish or will breed under such drastic alterations to their natural way of life, although all young mammals can be tamed when nurtured under the right conditions.

A young animal if removed from its mother and its natural environment to be reared in captivity, for example in a zoo, will, if it survives, adapt to its new way of life. This adaptation will include changes in its physical anatomy and in its behaviour. Some of these changes are obvious, some are barely perceptible, but if the animal breeds in captivity they will be more marked in the next generation (see below). The mammalian body is a much more plastic and changeable structure than might be thought, and even the skull and skeleton of a tame animal may exhibit distinct differences from that of its wild parent.

Future generations of the tamed animal, whether these lived in the prehistoric period or at the present day, will be subjected not primarily to natural selection but to artificial selection by man for characters that may be favoured for economic, cultural, or aesthetic reasons rather than for survival of the species. Over a long period of selective breeding dramatic alterations may occur in the appearance of the domesticated animal although these will be always constrained by genetic barriers. It is for this reason that the effects of taming and domestication produce the same general physical changes in widely different groups of mammals. Many of these changes originate from the retention of juvenile characters into adult life rather than from genetic alteration and they therefore exhibit the same form in mammals as different as the pig and the dog. Examples of such changes are the deposition of fat under the skin, the shortening of the jaws, and the curled tail, all of which will be discussed in further detail, later in this chapter.

Domestication causes an imbalance and disruption in the rate of growth of different parts of the organism, resulting in morphological proportions in the adult animal that differ from those of its wild counterpart. The mechanism of how this happens is not well understood but it is likely to be due to stress and to hormonal changes as a result of the animal's emotional and physical dependence on man. Domesticated animals that return to living in the wild, that is they become feral, will usually revert by natural selection to a physical form that is closer to the wild species as a consequence of the loss of this dependence. The present-day feral pigs of New Zealand are a good example of a population of animals that after a relatively short period conformed to a uniform type more closely resembling the wild boar than their highly domesticated antecedents.

The general effects of domestication can be summarized as follows:

SIZE OF BODY

The early stages of domestication of any species of mammal are almost always accompanied by a reduction in size of the body. This is so generally true that it is used as the main criterion to distinguish the skeletal remains of domestic from wild animals when these are retrieved by archaeological excavation of early prehistoric sites.

During the later stages of domestication animals that are either very much larger or very much smaller than the wild progenitor are selected for and breeds developed from them, for example, shire horses and Shetland ponies.

OUTWARD APPEARANCE AND PELAGE CHARACTERS

During the late Mesolithic and early Neolithic periods when man was experimenting with the breeding of early domestic dogs and livestock animals it is probable that those animals that were easy to distinguish and markedly different from the wild species would be especially favoured; they could be identified easily and they would increase the prestige of their owner (Fig. 1.1, p. 18). Thus a puppy that had a curled-up tail, or lop ears, or more white on its body than was usual would be nurtured more carefully and better fed than its litter mates. In this way greater diversity would be introduced into the stock of domestic animals. A disproportionate lengthening of the ears has occurred in most common domestic mammals with the exception of the horse, whilst an extra long tail is found in many breeds of sheep. A tightly-curled tail is common in many breeds of dog, occasionally in the cat, and in many pigs (Fig. 1.2).

Figure 1.2

Great variation is found in the wool and hair coats of most domestic animals, this being related to the climate where the breed has been developed as much as to man's preference. For example dogs and sheep from the alpine regions have extraordinarily thick woolly coats whilst those bred in the tropics may be almost hairless. In domestic long-woolled sheep the character for self-shedding with the onset of summer has been lost and the wool is continuously growing which means that, although the

Figure 1.3   Fat-tailed sheep, Iraq (photo author).

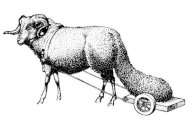

Figure 1.4

fleece has to be shorn by a shepherd each year, none of the wool is lost.

In the horses the hair of the mane and tail is much elongated; in the wild horse and in asses and zebras the mane is short and stands erect rather than falling to one side as in all breeds of domestic horse.

INTERNAL CHARACTERS AND DENTITION

There is a general tendency in well-fed domestic mammals for there to be a layer of fat under the skin and bands of fat through the muscle; in the past this was particularly favoured in beef and mutton. In wild animals surplus fat is usually stored around organs such as the kidneys rather than through the muscle. Sheep and pigs are prone to the development of excessive amounts of fat and in the sheep an anciently evolved character is the fat tail (Fig. 1.3). This was described by Herodotus* in about 450 BC, as follows (Fig. 1.4):

There are also in Arabia two kinds of sheep worthy of admiration, the like of which is nowhere else to be seen; the one kind has long tails, not less than three cubits in length, which, if they were allowed to trail on the ground, would be bruised and fall into sores. As it is, all the shepherds know enough of carpentering to make little trucks for their sheep's tails. The trucks are placed under the tails, each sheep having one to himself, and the tails are then tied down upon them. The other kind has a broad tail, which is a cubit across sometimes.   (III, 113)

In most domestic mammals the size of the brain becomes smaller relative to the size of the body and the sense organs become reduced. It is, however, in the skull, and to a lesser extent in the skeleton that the greatest changes may be observed.

* Translated by Rawlinson (1964, 1, p. 264).

Within a very few generations of breeding in captivity the facial region of the skull and the jaws become shortened, this being common in many species but is most apparent in early domestic dogs. At first there is no corresponding reduction in size of the cheek teeth which are genetically much more stable than the bones of the skull. This causes a crowding or compaction of the premolars and molars, a character that is used to distinguish the remains of the earliest domestic dogs from those of wild wolves.

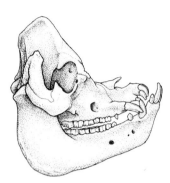

Figure 1.5   Skull of a pig of the Middle white breed, length *c*. 37 cm

Figure 1.6   Skull of an ox of the Niatu breed, Buenos Aires (BM(NH), 1887), length *c*. 36 cm

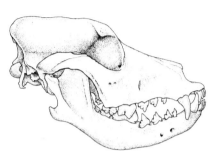

Figure 1.7   Skull of a Great Dane, length *c*. 26 cm

Shortening of the jaws and facial region is an example of the retention of a juvenile, or even foetal, character in the fully grown animal. As well as being common in dogs it is also found, sometimes to an exaggerated extent, in breeds of pig such as the Middle white (Fig. 1.5), and occasionally in cattle such as the South American breed known as Niatu (which may now be extinct, Fig. 1.6).

Following the reduction in size of the jaws, especially in the dog, there is a reduction in the size of the teeth which becomes permanent in the domesticated animal so that, for example in the Great Dane (Fig. 1.7), a breed of dog that is much larger than its progenitor the wolf, the teeth are still considerably smaller and have a less complicated cusp pattern than is found in the wolf. Similarly the tympanic bulla (bony case of the ear drum) of the dog's skull is considerably smaller than in the wolf (Fig. 1.8).

The horns of cattle, sheep, and goats show great diversity under domestication and this is often reflected in changes to the skull. The horn consists of a keratinous sheath overlying a bone core which is an out-growth of the frontal bone. Excessively large or long horns, as in the Ankole cattle of East Africa or the lyre-horned cattle of Ancient Egypt, have been presumably selected for aesthetic or ritual reasons and for the enhancement of status within the human community that has bred them.

Changes in the skeleton of domestic mammals are not often so marked as in the skull and are usually only related to alteration

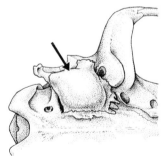

Figure 1.8   The tympanic bulla (arrowed) in the skull of a Great Dane (above) and a wolf. Bulla of wolf 30 mm long

in overall size of the body; however the muscle ridges and facets for articulation of the joints are reduced in domestic animals that are not much exercised, whereas in draught animals such as the ox or the horse they may be exaggerated.

BEHAVIOUR

With the exception of the domestic cat, all domestic mammals are derived from wild species that are social rather than solitary in their behaviour. As with certain physiological characteristics such as gestation period, blood group, and chromosome number, the structure of the social behavioural patterns of domestic animals is probably little changed from those that evolved in their wild progenitors, so that for example however different a Pekingese dog may look from a wolf its behaviour is still recognizably wolf-like. Any changes in behaviour that do occur appear to be a result of the retention of the juvenile, submissive behaviour of the young animal to its parent. This is obviously of great advantage to the human owner who wishes to retain dominance over the animal and it is probable that it has been highly selected for, and is correlated with the retention of juvenile features in the anatomy (Fig. 1.9, p. 35 and Fig. 1.10).

These aspects of behaviour in domestic animals have received little attention from ethologists until recently, perhaps because the wild parent species is so often either extinct or restricted to the most inaccessible refuges. However the work of Geist (1971, 1975) on the wild sheep of North America and of Schaller (1977) on the wild sheep of Asia, together with the numerous studies that have been carried out on the behaviour of the wolf will help to provide new insight into the social behavioural patterns of domestic animals.

CASTRATION

The behaviour of all domestic animals is drastically altered by the removal of the testes in the male (castration), and to a lesser extent by the surgical removal of the ovaries in the female, known as spaying. It is probable that at least the former operation has been carried out since the beginning of livestock husbandry for castration is an essential aid in the human control of animals. As is well known it turns an unmanageable, aggressive bull into a placid and submissive, draught ox, whilst a yowling, fighting tom cat becomes a fat and contented pet.

The physical changes brought about by castration are pronounced even if they are not as dramatic as the behavioural. After a young male animal has had its testes removed changes occur in its growth pattern so that its bones continue to grow in length rather than in girth, and more fat is laid down in the body tissues. In a castrated ox, for example, the body will be fatter than in a bull fed on the same diet but the limbs and the horns will be longer and more slender (Armitage & Clutton-Brock, 1976).

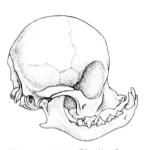

Figure 1.10   Skull of Pekingese, length *c*. 8 cm

2 Selective breeding and the definition of a breed

A *breed* is a group of animals that has been selected by man to possess a uniform appearance that is inheritable and distinguishes it from other groups of animals within the same species. It is a product of artificial choice of characters that are not necessarily strategies for survival but are favoured by man for economic, aesthetic, or ritual reasons, or because they increase the social status of the owner of the animals. These characters are usually short-lived and relate more to outward appearance and temperament than to the anatomical and physiological changes brought about in a population of animals during the long course of evolution. Unless they are extravagantly developed like the skull of the Pekingese dog the characteristic features that distinguish one breed of domestic animals from another will not be reflected in the skeleton. It is therefore always difficult and sometimes impossible to trace the history of a particular breed from the archaeological record.

Selective breeding was probably practised by the earliest Neolithic farmers for animals that were distinctive and submissive, as well as small, hardy, and easy to feed. Apart from the occasional figurine and the very rare scraps of horn or hide, the only way of assessing the results of selective breeding in the earliest periods is from the fragmentary bones and teeth of animals eaten for food and later retrieved by archaeological excavation. From later periods there is evidence from pictorial representations of animals and from these it can be seen that both the Babylonian and the Ancient Egyptian civilizations had developed definitive breeds of dogs, cattle, and sheep by the beginning of the second millennium BC (Fig. 2.1). In northern Europe there is little evidence for the presence of breeds in any of the farm animals until the Roman period when both the written record and the animal remains show that the Romans made positive efforts to 'improve' their stock. Perhaps they were the first Europeans to understand how to manipulate the natural diversity found within a single population of primitive domestic animals by the selection and reproductive isolation of favoured individuals, and this must have involved some understanding of the mechanisms of inheritance.

From the earliest post-Pleistocene archaeological sites – those of the Mesolithic and Pre-pottery Neolithic periods (about 8000 years ago, see chart, Appendix II) – it is not unusual to find the skeletal remains of mammals, often in a very fragmentary condition, that are difficult to describe because it cannot be determined whether they represent animals that were wild, tamed, or in any other way exploited by man during life. In the past these specimens were often classified as palaeontological 'fossils' and given new names according to the formal zoological code, whereas nowadays they are more usually treated as evidence of incipient domestication. The precise naming of these finds still presents difficulties as does the naming of the later stages in the sequence from wild progenitor to the highly specialized animals that live in association with man today. For this reason the meanings of some of the common terms used in biological classification are discussed here.

Classification is the process of establishing and defining systematic groups of organisms. The groups are known as *taxa* (singular: *taxon*), and nomenclature is the allocation of names to these taxa. The formal nomenclature of an organism, as opposed to its vernacular or common name, is governed by fixed rules that for animals are contained in the *International Code of Zoological Nomenclature*.

The unit on which biological classification is based is the *species*, usually defined as a group of actually or potentially interbreeding natural populations that are reproductively isolated from other such groups. Within the species there is always a degree of

Figure 2.1 Hunting of wild animals and the husbanding of domestic livestock from Tomb 3, Beni Hasan, Egypt, *c.* 1900 BC. Note the spotted, hornless cattle together with those of the lyre-horned breed and also the different kinds of dogs (from Newberry, 1893).

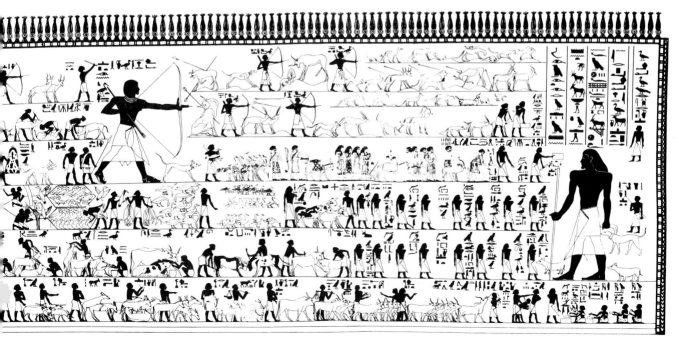

variation and diversity reflecting the unique character of each individual as well as the overall differences that may be seen in local populations. All species vary, both in space and in a temporal sequence. Wild animals vary in space, for example, according to climatic conditions as well as within local populations that may be either temporarily or permanently, reproductively isolated. In time they vary in the course of evolution, which may be sometimes very fast or may take millions of years. Every individual animal recognizes its own kind and as general rule will breed most readily with a member of its kindred group.

The terms subspecies, breed, variety and race are used to describe the diversity found throughout the animal kingdom below the level of species. A *subspecies* is a distinctive, geographial segment of a species, that is, it comprises a group of wild animals that is geographically and morphologically separate from other such groups within a single species. It is the lowest category of classification that can be included within the formal, Linnaean name of the species. This is achieved by giving it a third Latin name or trinomial as in *Felis silvestris grampia*, the Scottish wild cat, where *Felis* denotes the genus, *Felis silvestris* the species, and *grampia* the subspecies (Fig. 2.2).

Figure 2.2   Scottish wild cat, *Felis silvestris grampia* (photo Geoffrey Kinns).

With living domestic animals the variation that occurs within the species is described in terms of *breeds* rather than subspecies. The breed and the subspecies, as categories of classification, are, however, sometimes confused with each other and even used interchangeably which is occasionally justified but more often results from a misunderstanding of the concepts of taxonomy. The difference between the two terms is that a subspecies is always restricted to a given locality where it has evolved as a result of reproductive isolation, whereas a breed is a product of artificial selection by man, and geographical barriers need play no part in its development. An example of a breed of this kind can be seen in the Apis bulls of Ancient Egypt. The birthplaces of Apis bulls were dispersed throughout Egypt but the breed was well-defined; the animals were black with white spots and a white saddle patch and they had to have a black spot on the tongue. The cult of Apis lasted for more than a thousand years, the bulls being bred purely for religious reasons and when they died they were buried with great ceremony in an underground rock tomb at Saqqara.

Figure 2.3   Detail from Fig. 2.1.

Another example of a breed that is not restricted to a geographical region is the ancient running-hound or greyhound that was bred for hunting throughout Europe and Asia (Fig. 2.3). There are, however, other groups of domestic animals, particularly livestock, that are intermediate between a breed and a subspecies. In these, geographical isolation and adaptation to a harsh environment have formed a population of animals into a discrete entity. An example of this can be seen in the primitive domestic sheep from the islands of St. Kilda in the Outer Hebrides of Scotland, known as Soay sheep from the name of one of the islands on which the sheep were restricted until 1932 (Fig. 2.4, p. 35). This population of sheep has undergone little artificial selection since it was first taken to St. Kilda, probably during the Neolithic period, but has evolved into a uniform group adapted for survival in the tough conditions provided by a small island in the North Atlantic and a peasant community always on the borderline of poverty. These sheep are a relic of prehistoric farming practice and they can be used as a model to exemplify how the first domestic animals were moulded by, and integrated into new environments that man had only partially under his control.

Probably the ancient breeds of livestock on the mainland of Britain, particularly the sheep and cattle, were also moulded in this way so that they became adapted to the varying conditions prevailing in quite small-scale local environments. The separate status of the long-established breeds of cattle is lent support by the work of Jamieson who in 1966 tested the proteins in blood serum, called transferrins, in 63 breeds of domestic cattle with special attention being paid to those from Britain where the blood sera were examined from 21 breeds. The data obtained by Jamieson, which are quoted in Figure 2.5, show the genetic

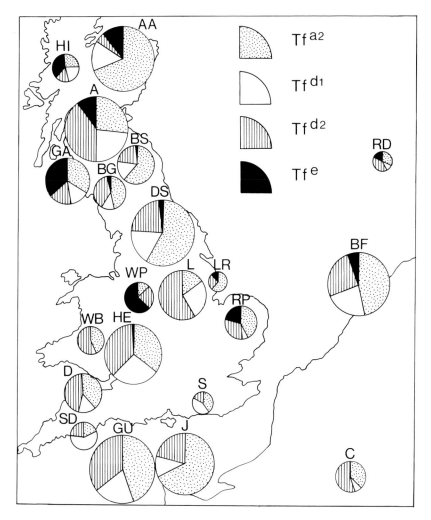

KEY
Cattle breeds represented by
their initial letters:
A   Ayrshire
AA   Aberdeen Angus
BF   British Friesian
BG   Belted Galloway
BS   Beef Shorthorn
WP   White Park
C   Charollais
D   Devon
DS   Dairy Shorthorn
GA   Galloway
GU   Guernsey
HE   Hereford
HI   Highland
J   Jersey
L   Longhorn
LR   Lincoln Red
RD   Red Dane
RP   Red Poll
S   Sussex
SD   South Devon
WB   Welsh Black

Figure 2.5 Geographic
distribution of allelomorphic
genes at the *Tf* locus in 21
breeds of cattle in England
and adjacent countries. The
positions of the circles on the
map indicate the probable
geographic origins of the
breeds whilst their size is
proportional to the numbers
of cattle tested (from
Jamieson, 1966).

variation in the breeds and throw some light on their geographic
origins that for some date perhaps from the prehistoric period.

It would be wrong, however, to suggest that breeds of cattle
should be given subspecific names on the grounds that they
constitute discrete geographical units, especially as with present-
day farming techniques they are most unlikely to remain isolated,
and for many breeds their vernacular names already precisely
describe their place of origin.

Not all variation within a species can be divided into discrete
segments. Often the different populations will merge into each
other and when this occurs across separate geographical regions
the subspecies are said to form a *cline*. A cline is not a taxonomic
unit but a gradation of measurable characters that are unidirec-
tional and pass into each other without discrete breaks in
sequence. Both breeds of domestic animals and subspecies of
wild ones can form clines and both develop as adaptations to
local environments and micro-climates. For example, the sub-

species of *Felis silvestris*, the wild cat, form a geographical cline over Europe and parts of Africa and Asia. This can be paralleled by the clinal distribution of the domestic cat (descended from *Felis silvestris*) where different coat colours and markings are found more commonly in some regions than in others. The map (Fig. 2.6) shows the relative frequency of blotched tabby cats over Europe. This kind of marking has arisen as a highly successful mutant from the wild type of tabby which is more closely striped and spotted (Fig. 2.7, p. 35). It is particularly common in the

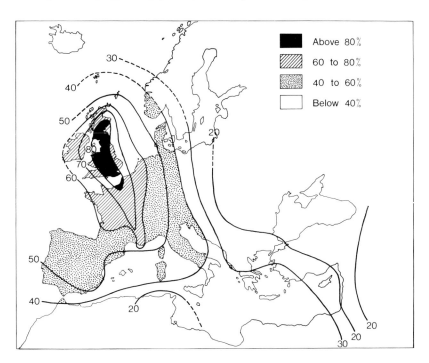

Figure 2.6   The density and distribution of blotched tabby cats over Europe and western Asia. The figures on the map are percentages. The highest densities are found in the largest cities (after Todd, 1977).

urban rather than the rural environment so that this cline may be said to run along an urban-rural axis (see also Chapter 10).

Below the level of subspecies and breed there are various terms for describing the diversity within populations of animals such as *variety*, *race*, and *strain* but their lack of precise definition can lead to confusion and they are better avoided. The term variety has several meanings: it can be applied to a discrete population within a species when it would mean much the same as a subspecies, or it can be used for an individual variant, or for a group of variants that are associated together and breed with each other but do not form a separate population. This last application is using the term variety as a synonym for *race*. Another term which is used rather vaguely and is also synonymous with race is *form*.

These terms can sometimes be useful in general description, as for example in 'the races of man', but they have no validity within the formal nomenclature of a species of animal and should never be given trinomials. (This is not necessarily so with cultivated plants for which there is a separate code of nomenclature.)

Unfortunately in the past and occasionally even now the skeletal remains of domesticated animals from archaeological sites have been ascribed to subspecies and at the same time described in terms of modern breeds. The standard example of this is to be found in the classification by Studer (1901) of the skulls of domestic dogs from the early prehistoric period. These specimens which are often in a very fragmentary condition and may be found as an isolated 'fossil' have been given subspecies and breed names according to the relative basal lengths of the skull. This system has been repeated time and again in the literature with extraordinarily little thought for its meaning and implications; it was even reiterated by Zeuner as late as 1963 in his renowned book, *A History of Domesticated Animals*. Thus a dog skull with a basal length of 171–189 mm was named *Canis familiaris matris-optimae* (Fig. 2.8) and described as a primitive sheepdog, whilst another skull from a quite different part of the world with a basal length of 164 mm was ascribed to *Canis familiaris intermedius* and was claimed to be the ancestor of various breeds of hound.

Figure 2.8   Skull of a canid from the Natufian levels of the cave of Wady el-Mughara, Mount Carmel, and named by Bate (1937) *Canis familiaris matris-optimae*.

It is true that the naming of archaeozoological specimens such as these dog skulls poses several difficult problems but they are only compounded by dealing with them in this manner. There can be no justification for aligning the skulls either with present-day breeds of domestic dog or with subspecies of wild animals and this applies to all such finds. Furthermore to assign trinomials to animal remains from any archaeological period on the basis of single characters that are unlikely to be unique in one group is to contradict all the principles of taxonomic classification.

Although it is certainly true that the Ancient Egyptians had definitive breeds of dogs, cattle, and sheep it is necessary to remember when tracing the history of domestication how ignorant people were in the past of the laws that govern the inheritance of variable characters. Nowadays every literate person who is concerned with the breeding of animals for economic profit, or with their history from an academic point of view, has at least an elementary knowledge of the concepts of evolution, and natural and artificial selection, but in the past there was very little understanding of how these processes worked. Inevitably therefore the production and retention of favoured characters through several generations was very much a hit and miss affair and remained so from the time of the Ancient Egyptians until after the publication of *The Origin of Species* by Darwin, in 1859.

Furthermore it is probable that many modern breeds of domestic animal, for example the greyhound, that appear to resemble ancient ones have no pure-line ancestry but result from manipulation by artificial selection of the natural diversity present in the species. This can produce breeds with a similar appearance in separate countries and at different periods without there being any close relationship between them.

With a few exceptions such as the pure line of Pekingese dogs bred in the Imperial Palace at Peking over hundreds of years, and the Arabian horses from central Arabia the production of well-defined breeds of domestic animals was not in the past controlled in the rigid way that we are used to. Nowadays the pedigrees of individual animals are carefully documented and the conformation of each creature is scored against standards set by a show ring judge or a breed society. All this is an innovation of the last hundred years. It has happened because of the general increase in knowledge about the rules of inheritance and evolution, and also because there are many people in affluent societies today who have the urge to raise their status by owning the perfect example of a breed, be it a stud bull or a Pekingese dog.

Before this century the development, or it could be called evolution, of a breed was a slow process that was in part a result of artificial selection but was also a response to the local environment, so that it was not unreasonable to think of, say, a breed of cattle such as the South Devon as a geographically isolated unit and therefore akin to a subspecies. This is no longer true, however, because the specialization that is inherent in modern farming, including the widespread use of artificial insemination, has changed the process. Cross-breeding now occurs on a world-wide basis and the environment that the new livestock breeds are often best adapted to is the artificial world of the factory farm. Therefore in giving the definition of a breed as in the first paragraph of this chapter it is necessary to add that a breed differs from a subspecies in that it is not nowadays restricted to a geographical locality.

NOTE The system of nomenclature used for the domestic mammals and their wild progenitors throughout this book is explained in Appendix I.

3 Dogs

Jackal, *Canis aureus*

Coyote, *Canis latrans*

ORDER CARNIVORA, FAMILY CANIDAE

The increase in the world's population of dogs matches that of mankind for there are now dogs in every part of the world that is inhabited by man and today there are more than 400 breeds of domestic dog. Each of these breeds owes its existence to artificial selection by man, because every dog, whether it is a Great Dane or a Chihuahua, is the descendant of wolves that were tamed by human hunters in the prehistoric period. The process of taming probably began at least 12 000 years ago but how much has changed in the actual relationship between man and dog in that period it is difficult to assess. It may be that in fact there is very little difference and the relationship is much the same now as it was at the end of the Ice Age. This is because the remarkable kinship and powers of communication that exist between human beings and dogs today have developed as an integral part of the hunting ancestry of ourselves and the wolf. It is a biological link based on social structures and behaviour patterns that are closely similar because they evolved in both species in response to the needs of a hunting team, but which endure today and have become adapted to life in sophisticated, industrial societies.

Dhole, *Cuon alpinus*

At the end of the Ice Age in an environment where wolf and man were competing for the same food it is not difficult to surmise how an alliance could be formed between them. The women and children of the hunting communities would give succour to any animal that would stay near them and young canids would be tamed along with many other animals, both carnivores and ungulates, the species depending on the locality. In Europe the canids would be the wolf with jackal in the east, in North America the wolf and coyote, in South America various species of 'fox' and the bush dog, in Africa the jackals and perhaps the hunting dog, and in Asia the wolf, jackal and dhole (Fig. 3.1). Most of these associations would be ephemeral, if the tamed animals lived to be adults they would move off to find their own food, live on their own and find their own mates. This is because they were solitary animals by nature without the complicated social behaviour patterns that are found in only the two species of wolf and man. As with human communities the

Bush dog, *Speothos venaticus*

Fox, *Vulpes vulpes*

Figure 3.1

34

Figure 2.7   Striped, domestic tabby cat in Iraq with markings that fairly closely resemble those of the wild cat *Felis silvestris libyca*, see p. 31 (photo author).

Figure 1.9   Pekingese, see p. 25 (photo G. Napier).

Figure 2.4   Soay sheep on the island of Hirta, St. Kilda, see p. 29 (photo P. A. Jewell).

Figure 4.2   Bushman woman carrying a child on a bag of mongongo nuts, see p. 47 (photo Lawrence K. Marshall).

Figure 3.3 'Smiling' mouth of a contented dog (photo P. A. Jewell).

social structure of the wolf is based on a hierarchy of dominant and submissive individuals who are constantly aware of their status in respect to each other. This is the basis of what we interpret as the reciprocal affection that is apparent between man and wolf, or man and dog.

Many other species of animal also depend on a leader but their overall social structure does not parallel that of man in the close way as does that of the wolf. For example, amongst the canids the hunting dog, *Lycaon pictus* (Fig. 3.2), is a most highly social carnivore that lives and hunts in packs but the dominance hierarchies in this species are not well developed and the social behaviour is more dependent on the mutual regurgitation of food and less on communication by facial expression and posture than in the wolf. So that if man is not prepared to receive the regurgitated offerings of a hunting dog into his own mouth his powers of communication with the animal are going to be limited. Whereas there can be such a close empathy between man and dog that if a puppy is reared amongst a human family that smiles a lot the dog will actually mimic this expression of pleasure by a sideways grin of the lips and muscles around the mouth (Fig. 3.3). This particular facial expression is never seen in wild wolves.

Figure 3.2 African hunting dog, *Lycaon pictus*

It can be seen how young wolf cubs could be drawn into human hunting communities towards the end of the Ice Age. Sometimes they would be eaten or would die or run off but occasionally a particularly placid or submissive puppy would survive to become an adult wolf that would accept the human group as its pack – it would become socialized or tamed for the duration of its life. Within recent years many people have shown that it is relatively easy to hand-rear wolf pups and to keep them in a tamed condition as adults, and of course the Eskimos in Canada have always known this and have not only crossed wolves with their own dogs to improve their stamina but have used wolves in their teams of sledge animals. There is, however, a great difference between a tamed wolf and a domestic dog.

Domesticated wolves would have been a natural development from pups that were caught in the wild and reared in a human community, but only after the descendants of these wild pups had been bred in captivity over several generations. A domesticated animal is always the product of a breeding population that is isolated from the wild species. The wild temperament has to be bred out of it which means that the animal has to lose the tendency to react in a defensive way to the unfamiliar, whether this is in the form of a person that it does not know or different territory, or any other traumatic experience. To have the potential for domestication the animal must be placid by temperament which means it must be submissive but it must not be over-fearful. Within any litter of wolf pups there will be wide range of different temperaments just as there will be in a family of children and in addition some will be weaklings and some will be especially hardy. Just occasionally a pup would have the correct combination of physique and temperament to make the necessary

Figure 3.4 Australian Aborigines with their tamed dingoes (*Canis dingo*). Although dingoes have lived wild in Australia for thousands of years they are not indigenous wild animals but are descended from domestic dogs that travelled to the continent from south-eastern Asia with the early human immigrants. In their morphology dingoes are closely related to the Indian pariah and New Guinea dogs. They are most interesting relics of the dogs that must have been widespread throughout western and eastern Asia during the prehistoric period and as such they should be conserved (photo *Illustrated London News*, 1937).

adjustments that would enable it to survive in a human community, live to be an adult, and even to breed. Because the offspring of such a wolf would be protected by man and could obtain their food by scavenging rather than by the more arduous and tough life of the hunter of large game, those that differed slightly from the rest of the wolf species could survive. Indeed they were probably given preferential treatment by their owners because they looked distinctive, they had curled-up tails, or were black and white instead of the normal wolf colour, or they had large eyes, or were in some other way more endearing than the rest of the litter of pups. These characters were inherited and so in time a separate kind of animal evolved, the *dog* (Fig. 3.4).

Columella, the Roman authority on agriculture, described in the first century AD why it was advisable to breed dogs that were different in appearance from the wolf* (Fig. 3.5):

Figure 3.5   Mosaic from Pompeii (photo The Mansell Collection)

* Translated by Forster and Heffner (1968, p. 307).

As guardian of the farm a dog should be chosen which is of ample bulk with a loud and sonorous bark in order that it may terrify the malefactor, first because he hears it and then because he sees it; indeed, sometimes without being even seen it puts to flight the crafty plotter merely by the terror which its growling inspires. It should be the same colour all over, white being the colour which should rather be chosen for a sheep-dog and black for a farmyard dog; for a dog of varied colouring is not to be recommended for either purpose. The shepherd prefers a white dog because it is unlike a wild beast, and sometimes a plain means of distinction is required in the dogs when one is driving off wolves in the obscurity of early morning or even at dusk, lest one strike a dog instead of a wild beast. The farmyard dog, which is pitted against the wicked wiles of men, if the thief approaches in the clear light of day, has a more alarming appearance if it is black . . .   (XII, 3)

During the Roman period dogs were already as highly domesticated as they are today although the diversity of breeds was not nearly so great. It is probable, however, that some pure lines were already differentiated and, for example, it is likely that dogs resembling the present-day Pekingese were already being bred in China.

There are many breeds of dog that appear to bear no resemblance whatsoever to the wolf and perhaps we should consider the creation of such an animal as the Pekingese to be one of man's greatest achievements though it took, maybe, thousands of years of selective breeding to produce. But although the outward appearance of the Pekingese is so dramatically different from that of the wolf its anatomy and physiology are still remarkably similar. The general morphology of the bones, apart from their proportions, are the same, so is the dentition, and the shape of the brain and the gut. The food of the Pekingese is digested in the same way as the wolf's, it has the same internal and external parasites and its young take the same length of time to develop in the uterus. Also, despite the rather absurd and babyish look of this little dog its behaviour can still be remarkably wolf-like. The word babyish is actually the clue to the way in which the wolf has been transformed into this other creature because it is the retention of juvenile or even foetal characters in the adult animal that is responsible for the change. The very short facial region of the skull, large brain case, big eyes, short legs, curly tail, and soft fur are all characters that are found in the foetal wolf but in the wild animal are lost either before birth or before the period of growth is completed.

The domestic dog has one capability that is never well developed in any canid living wild, and this is the propensity to bark. Wolves and coyotes will bark occasionally in the wild and other members of the canid family will learn to vocalize in captivity but the deep bark of the large breeds of dog, the baying of the Bloodhound, and the yapping of the toy dogs is a product of domestication. A few breeds such as the primitive Basenji do not bark although they can learn to do so, but most dogs have a low level of arousal and it requires little incentive to start them off. As it is obviously useful for a guard dog to bark at strangers it is most probable that this form of vocalization has been highly selected for. In addition some dogs may bark in an effort to communicate with humans and it could be that they are even attempting to mimic the human voice. The bark is an attention-seeking device that can be associated with play-soliciting, hunting, or aggression.

Although the wolf does not normally bark in the wild it can learn to do so in captivity and in general the vocalization patterns of the wolf and the dog are very similar. Those of the jackal, however, are quite different and provide sound evidence that the jackal has had little to do with the ancestry of the dog.

In 1950 J. P. Scott, who carried out fundamental investigations in the U.S.A. into the social behaviour of the wolf and how it relates to socialization in the dog, wrote that the patterns of behaviour of dogs in human society are the same as those of wolves in wolf society. Since that time much further work has been carried out on behaviour patterns in wild and tame wolves, in dogs, and in people, so that it is now possible to state that wolves also behave in human society as they do in wolf society.

Many social animals depend on a hierarchical system of differentiation between individuals with reference to reproduction and maintenance of territory, but it is only in the large carnivores and in some primates that there is such a rigid structure of personal relationships between individuals as is found in the wolf and in man. For this to occur there has to be not only dominance behaviour but also submissive behaviour. Submission has been defined by Schenkel (1967), for the wolf and the dog, as an impulse and effort on the part of the inferior animal towards friendly and harmonic social integration. It is ritualized behaviour that is characterized by the combination of inferiority and a positive social tendency ('love') and it does not contain any element of hostility. The form that the submissive behaviour will take depends on the attitude and behaviour of the superior individual. If he responds in a negative or intolerant way the inferior will not persist but if the superior individual wishes to enter into friendly communication with the inferior then harmonic social integration can take place. Active submissive behaviour in canids is derived from the role of the pup in relation to its parent and is centred around begging for food by the cub, olfactory (smelling) investigation, and ano-genital licking by the mother. Schenkel illustrated a typical form of submissive behaviour by the following observation on a pack of wolves in Whipsnade Zoo, England:

After rising in the morning the leader of the pack walked around sniffing the soil. Somewhere he stopped and dug out a big bone. He seized it and passed, the bone in his jaws, near the pack in 'proud' gait with his tail up. Now the other wolves rose, surrounded the

a

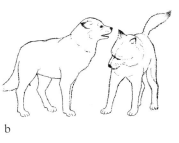

b

c

Figure 3.6 Communication between submissive and dominant dogs, see text. The dominant dog has its tail raised (after Schenkel, 1967).

Figure 3.7 Ritual submissive behaviour (after Schenkel, 1967).

leader, and joined in a begging ceremony. The leader first growled and continued his walk. Then he dropped the bone and left it. The others surrounded the bone for an instant, and then left the spot. Obviously the begging-for-food-scene was not a real but a symbolic one. The bone was only a requisite for the leader and the pack to join in a ceremony of harmonic social integration.

The different types of submissive behaviour depend on the nuances in attitude of the superior animal. The more inquisitive and severe the superior is, the more the inferior tends towards the passive type of submission, but if the superior animal is tolerant without being dominating the inferior may try to assert himself. On the other hand if the superior is aggressive the submissive animal will retreat. This ritual behaviour can be seen in any two dogs that meet on a 'walk' in a suburban park, any day; it can also be observed in wild or in tame wolves.

Figure 3.6a–c shows the ritual body positions of a submissive dog on meeting a more dominant individual, whilst Figure 3.7 shows the passive submission of a dog that allows its muzzle to be grasped by a superior animal. This behaviour can be observed repeatedly in dogs and wolves and although it is present in other canids it is not so well developed. The superior animal can always be distinguished from the inferior by the upward carriage of his tail, his upright ears, and 'proud' stance. It can be seen with breeds of domestic dogs that those with long pendulous ears like the spaniels have a permanently submissive appearance even to humans who describe the behaviour as 'fawning', whilst breeds that are used as police and guard dogs may have their ears clipped so that they are permanently upright, thus enhancing their dominant appearance. The dominant animal will often allow the submissive to approach the vulnerable side of his neck with open jaws or even to grasp his head in his mouth. This action was originally misinterpreted by Lorenz who believed that it was the dominant animal that was holding the submissive in a position of inhibited attack but Schenkel has shown how it is the reverse situation that is correct and although the inferior animal looks as though he can bite the other, in fact he does not dare to do so (Fig. 3.6b).

The value to the species of this dominant and submissive behaviour is that it inhibits fights-to-the-death between carnivores whose way of life is centred on the slaughter of large animals for food and who are therefore especially skilful at killing. There is of course much aggression amongst a pack of wild wolves and each individual has to maintain or improve his position in the hierarchy but serious intraspecific fighting is uncommon. It is the development of ritual, submissive behaviour that permits the wolves to live together in groups larger than the basic family unit. It is this type of behaviour that also allows humans to live together and it is the basis for the mutual understanding which is often apparent between even a very young child and a puppy.

We cannot know when human hunters first tamed wolves any

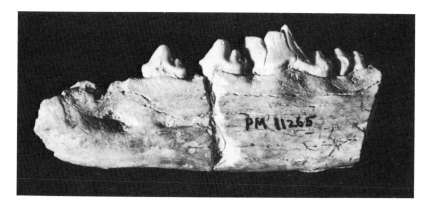

Figure 3.8   Left mandibular ramus of an early domestic dog from Palegawra Cave, Iraq, approx. 8 cm long (photo Field Museum of Natural History, Chicago).

more than we can know when they first killed an animal with a wooden spear or a stone projectile, because there is no trace of the animals left in the archaeological record. It is only when the association between man and wolf became a rather commonplace affair and the wolves began to differ in their physical characteristics from their wild forebears that their bones can be distinguished and they can be separated out from the rest of the food debris on an archaeological site. At present one of the earliest records of such dog remains comes from the late Palaeolithic cave of Palegawra in Iraq and is probably dated to about 12 000 years ago (Turnbull & Reed, 1974; Fig. 3.8). This mandible can be distinguished from that of a wild wolf by the relatively small size of the jawbone and the compaction of the teeth particularly in the premolar region. This is a typical feature of early domestic dogs.

The majority of the remains of the earliest domestic dogs have been retrieved from archaeological sites in western Asia, although small numbers have also been found in North and South America, north west Europe (England and Denmark), Russia and Japan. They are nowhere very common until the Neolithic period when livestock animals are of course also represented. It seems probable from these widely distributed but sparse finds that man tamed wolf pups in many parts of the world where both species were living as hunters of large game towards the end of the Pleistocene period and that therefore several sub-species of wolf have contributed to the ancestry of the dog. In western and southern Asia the subspecies would have been the Arabian and Indian wolves, *Canis lupus arabs*, and *Canis lupus pallipes*, whilst in Europe and North America it would have been the much larger wolf *Canis lupus lupus* that was adapted to life in a colder climate (Fig. 3.9). Although it appears reasonable to assume that these large northern wolves have contributed to the early ancestry of the dog there is very little evidence for it from the skeletal remains. All the finds from the upper Palaeolithic and Mesolithic periods that can be said with certainty to represent domestic dog are from rather small animals with teeth that are

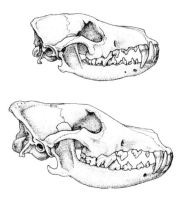

Figure 3.9   Skulls of the small Arabian wolf, *Canis lupus arabs*, above, and the much larger European wolf, *Canis lupus lupus*.

closer in size to the small Asiatic wolves than to the large northern subspecies. From the early Neolithic period there are some remains from what appear to be large dogs, notable amongst these are the 53 cranial and mandibular fragments from the site of Jarmo (*c.* 6600 BC) in the foothills of the Zagros mountains in northern Iraq, some of which have large teeth almost comparable in size with those of the European wolf whose range extends to this mountainous region. It is difficult to know for certain whether these remains represent wild or tamed wolves or large dogs. Another site that has provided canid remains that appear to be intermediate between dog and wolf is the settlement at Vlasac on the Danube at the Iron Gates gorge in Rumania. This site is very interesting because although it is rather late in date (*c.* 5400–4600 BC) it has a Mesolithic culture and there is no evidence for domestic animals other than the dog or for cultivated grain. This is probably because of the mountainous and isolated nature of the terrain which made it suitable only for a hunting and fishing community rather than for agriculture. A great many fragments of canid bone were retrieved from this site (1914 specimens) and most of these were from small domestic dogs some of which had been certainly eaten because the bones had been chopped. There were, however, a few jaw fragments and teeth which appear to be between wolf and dog in size and Bökönyi (1975) who has described this fauna has suggested that these specimens represent dogs that were domesticated *in situ* from local wolves, but such finds are very uncommon.

The dogs of these early periods, before the invention or widespread use of agriculture, were already quite variable in size and they probably also varied in their pelage, length of ears and tail, and shape of facial region. Some would have had a more marked 'stop' to the skull (i.e. a greater cranio-facial angle) than others and some would already have had the look of a greyhound, with a long muzzle and long rather fine-boned limbs (Fig. 3.10). As described in the chapter on breeds, however, it is not acceptable to divide these dog remains into separate categories or subspecies, let alone into breeds. Examination of a population of scavenging curs associated with any single peasant community at the present day would be likely to produce all the shapes and sizes of skull that have been found throughout the world from pre-Neolithic times.

That is not to say that local populations of prehistoric dogs did not differ from each other, only that it is inaccurate and inappropriate to describe them in terms of modern breeds. The only assessment that should be made from skeletal remains is one of size and proportions drawn from direct measurement of the bones and teeth. An example of how this is done can be quoted from the description of the complete skull and skeleton of a dog that has been excavated recently from the Neolithic flint mines at Grime's Graves in Norfolk, England. Detailed measure-

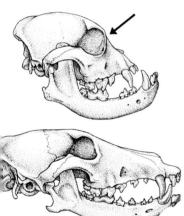

Figure 3.10   Skull of a short-faced domestic dog with a marked 'stop' arrowed, to compare with that of a greyhound.

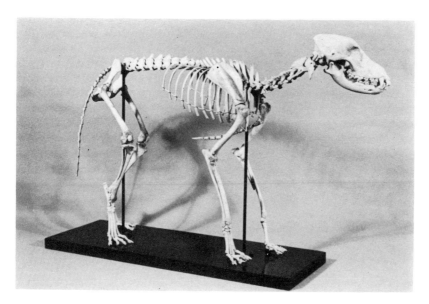

Figure 3.11 Skeleton of a Neolithic dog from Easton Down flint mine (photo Salisbury and South Wiltshire Museum).

ments were recorded for all the bones and teeth and compared with other remains of dogs from British sites of the same period, that is about 2000 BC. From this metrical data it can be shown that the skeleton from Grime's Graves is similar to the two equally complete skeletons from the nearly contemporary sites of Windmill Hill and Easton Down in Wiltshire, and that these dogs all had a shoulder height of about 50 cm (Burleigh *et al.*, 1977; Fig. 3.11). It would be rash to draw general conclusions on the appearance of dogs in Britain during the Neolithic from the evidence of so few skeletal remains but it can be tentatively assumed that there was less variation in the population at this period than in later times. In the Iron Age for example, the remains of dogs are much more variable whilst from Roman Britain there are remains of dogs that range in shoulder height from 23–72 cm (9–28 in.), see Harcourt (1974).

The Romans certainly had specific breeds of dog and both very large fighting dogs and tiny lap dogs are well recorded both in literature and art, and from osteological remains. Pliny writing in the first century AD had this to say of the lap dog.[*]

As touching the pretty little dogs that our dainty dames make so much of, called Melitaei in Latin, if they be ever and anon kept close unto the stomach, they ease the pain thereof.

The Romans were probably the first people to develop what we would call definitive breeds of dog in Europe, but from the pictorial evidence of earlier periods it is clear that there were many different shapes and sizes of dogs in Ancient Egypt and there were also large, heavy hunting dogs throughout western Asia from at least 2000 BC (Fig. 3.12). These dogs, which do look like present-day mastiffs, together with the Ancient Egyptian hunting dogs, which closely resemble present-day greyhounds,

[*] Translated by Philemon Holland in 1601 and selected for re-publication by Turner (1962, p. 316).

Figure 3.12 Mastiffs from the palace of Ashurbanipal, Nineveh (Iraq), now in the British Museum, c. 645 BC (photo BM).

appear to be two breeds that have continued more or less unchanged until the present day. It is difficult to determine, however, whether the mastiff and the greyhound are really breeds with an unbroken line of 4000 years or whether the genetic diversity inherent in the species causes similar characteristics to re-combine so that the same type of dog is bred in different regions and at different periods when selective breeding is carried out for the same need, this being to provide hunting, racing, and guard dogs.

4 The origins of domestic livestock
– Why bother to farm

With carnivorous animals if the prey becomes scarce, the predators will also decline in numbers, but with human hunters this has not been so. Man, unlike all other animals, since early in his evolution, has been able to adjust his behaviour and manipulate the environment so that he is able, not only to remain the master predator, but also to maintain a steady growth rate in population numbers. The storage of food to be eaten later when times are lean has been the pivot for the human population explosion of the last 10 000 years.

It is difficult to preserve meat; it soon rots and it is then thoroughly unpleasant to eat, if not actually lethal. Therefore if game animals are not readily available to be hunted and killed whenever meat is required it is better to store the meat as *livestock* rather than trying to keep it as dead flesh, even if this is smoked or dried. This is probably the most basic reason why the steadily increasing human populations of the early Holocene began to change from hunters to farmers, approximately 8000 years ago. There are, however, many complicated questions that surround the issue of what provided the impetus for this most dramatic alteration in the way of life of mankind, and why it happened when it did, and where.

For the last hundred years archaeologists and anthropologists have tried to account for the change in man's way of life from hunting to the cultivation of plants and the husbandry of livestock. Until recently it was generally believed that the change was due to a natural evolution and that the settled life of the farmer was easier and more comfortable than the austere struggle for survival that the hunter and his starving family were assumed to endure. But then in 1968 with the publication of the symposium volume *Man the Hunter* and the reports by Lee on the 'Kung Bushmen (1969) in the Kalahari desert of southern Africa, it seemed that most hunter-gatherer communities have a rather easy and trouble-free existence with plenty of food readily accessible to them. This appears to be true irrespective of whether they are Kalahari Bushmen, Arctic Eskimoes, or Australian Aborigines, as long as there is a balanced ecosystem and the people are not under subjugation to invading foreigners who with their more

Distribution of Hunter-gatherers

c. 10 000 years ago

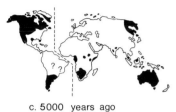

c. 5000 years ago

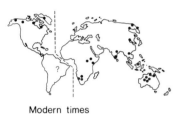

Modern times

Figure 4.1   (after Lee and De Vore, 1968).

sophisticated technical skills are able to eradicate the food supply. However, just as in the past it was tempting to simplify the change from hunter to farmer as a progression forwards from savagery to civilization, now we are probably extolling too highly the supposed easy life of the hunters. From studies of modern hunters, although all these live today in what may be termed refuge areas, it has become fashionable to extrapolate back to the end of the Pleistocene, before the invention of agriculture, and to suggest, as has been done by Sahlins (1972) that this was the period of the original affluent society.

If the subsistence level of the world's populations of hunter-gatherers was highly satisfactory at the end of the Pleistocene then a hypothesis has to be postulated to explain why the change over to the drudgery of farming took place at this time. For there is no doubt that it is much easier to gather wild plants and kill animals for food by hunting than it is to harvest primitive cereals, build fences to restrain livestock animals and be ever aware of the constant necessity to provide food and water for this walking larder. And yet it is a fact that whereas 10 000 years ago all human beings lived as hunter-gatherers, 5000 years later the majority of people in the most densely populated areas had become farmers (Fig. 4.1).

A recent comprehensive review by Cohen (1977) attempts to provide explanation of this transformation by means of the hypothesis of continuous growth in population. It would seem eminently reasonable to assume that if the cultivation of plants produces a greater quantity of food in a given area in a shorter time then the incentive to nurture the plants rather than to exploit them in the wild state will follow where greater productivity is required. In other words, a greater number of people can be supported from the cultivation of cereals than from wild plant and animal foods. There is nothing mystifying about agriculture and it is unlikely that any group of people at any time or any place in the world has been unaware of the connection between the seed and the plant, but if there is plenty of food around there is no need to go to the trouble of cultivating more. In the words of the 'Kung Bushman in reply to Lee's question on why he hadn't taken to agriculture, 'Why should we plant when there are so many mongongo nuts in the world?' (Lee, 1968 p. 33). Mongongo nuts (*Ricinodendron rautanenii*) provide a highly nutritious staple food for the Bushmen and it is my belief that if a Mesolithic man in Britain had been asked the same question about 10 000 years ago he would have given the same reply but in relation to the hazel nut which appears to have equalled the mongongo nut in abundance (Fig. 4.2, p. 35).

Changes in climate such as occurred at the end of the Pleistocene combined with a reduction in habitable land, together with a lowered density of wild game animals at least in part due to over-hunting, was surely the motive for the inception of agriculture and the husbandry of livestock. Increasing pressure on the available food supplies in many parts of the world during the early Holocene meant that new methods of obtaining food had to be resorted to, just as at the present day the soya bean steak may have to replace the sirloin.

Archaeologists have tended to concentrate their excavations in western Asia, certainly one of the 'hearths of agriculture', but there is now good evidence to indicate that at the same early period, in the eighth millennium BC, agriculture was also initiated in south east Asia and in the New World. There was, however, no cultivation of plants or domestication of animals in northern or western Europe at this period where, as I have suggested, the wild hazel nut provided plant food and the abundant large mammal fauna, with the red deer in particular, provided meat.

The Tell* of Jericho, lying 366 m below sea level, in the Jordan valley is a unique archaeological site that has provided evidence for a continuity of habitation for a timespan of almost 9000 years, from the Mesolithic period, or Natufian as it is called in the Near East, to the Byzantine (Fig. 4.3). The earliest levels of the Tell (excavated under the direction of the late Dame Kathleen Kenyon from 1952–8) cover the phase when cereal cultivation and animal domestication first began, and indeed the plant and animal remains from this period do show morphological changes that are characteristic of incipient domestication. The cultural period following the end of the Natufian, when the demise of the hunter-gatherers began, is known in western Asia as the Pre-pottery Neolithic A (see chart, Appendix II); it began about 8000 BC and lasted for a thousand years. At this time the settlement consisted of a village of hunters who tended a few cereal crops but had no domestic animals. It was a very large village covering an area of about four hectares (10 acres) and housing perhaps as many as 2000 people, who had solid mud-brick houses and stone defence walls but no pottery (Kenyon, 1957). The evidence for the cultivation of cereals is based on the identification by Hopf (1969) of grains of emmer wheat, *Triticum dicoccum*, and hulled two-row barley, *Hordeum distichum*. It is likely that these crops were watered by a primitive system of irrigation.

That the people of the Pre-pottery Neolithic A period at Jericho were still hunters is irrefutable. The animal remains retrieved from the excavation show that they lived mostly on the meat of gazelles but they also frequently ate the meat of the common fox, *Vulpes vulpes* (Clutton-Brock, 1979). It has been suggested by Legge (1972, 1977) that the gazelles at Jericho and other sites in the Near and Middle East were, if not domesticated, at least tamed by the early Neolithic people, but there is no evidence for this in the bones which show no difference from those of modern wild animals. I believe it is more likely that the gazelle were hunted in an organized way by means of drives and

Figure 4.3   Excavation of the Tell at Jericho (photo British School of Archaeology in Jerusalem).

* A 'tell' is a large artificial mound that marks the site of an ancient settlement. The level of the site is raised by the successive building of mud-brick houses one upon another, for mud-brick, once it decays, cannot be re-used like stone.

surrounds as described by Henry (1975) than that they were held in any form of captivity. Gazelle were clearly very common indeed in western Asia at this period as were red deer, *Cervus elaphus*, in Europe and there would have been no problem about hunting and killing as many animals as would be necessary to feed even as large a community as that of Jericho in the Pre-pottery Neolithic A. In addition to gazelle and fox there were many other species of mammal that could be hunted, including the wild ox, *Bos primigenius*, wild boar, wild ass, goats, sheep, hares and a multitude of small animals and plants that could be readily gathered. So that although the basic meat supply was provided by gazelle the economy was broadly based on a large number of plant and animal foods and this is characteristic of all the many sites that have now been excavated in western Asia from this and somewhat later periods.

The climate was changing, however, and becoming more arid so that it was necessary for the people to congregate more closely around any permanent source of water. These settled human communities were steadily increasing in size and the numbers of game animals were decreasing. Although pottery had not been invented these first farmers knew how to store food in clay-lined storage pits and when they moved from place to place it is likely that they took a few goats and sheep that had been tamed as young animals with them as a supply of living meat. They probably tried taking gazelle with them too but this would not have been so successful. It is much easier to keep goats and sheep and even cattle and pigs in captivity than it is deer or gazelle. These animals are not easy to manage, they are territorial in their behaviour and much less flexible in their feeding habits than goats and sheep, also they will not breed easily if kept in close confinement.

The next period at Jericho reflects the changing conditions and the inhabitants' responses to the need to exploit different sources of food. This period has been called the Pre-pottery Neolithic B and it began about 7000 BC (see chart, Appendix II). Although there was still no pottery this is the period when the practice of agriculture became consolidated in western Asia. There is increased evidence for the cultivation of cereals from many sites besides Jericho (Renfrew, 1973), and there is the first evidence from changes in the morphology of the bones of sheep and goats to indicate that man had begun the process of domestication. He may also have kept pigs and cattle in captivity at this period but there is little evidence for this from their remains.

Archaeological sites from which evidence for early domestication has been obtained are shown on the map (Fig. 4.4). There has been some competition amongst archaeologists excavating in western Asia to produce the earliest signs of domestication in plants and animals. The precise localities and dates at which domestication first appeared is, however, of small importance

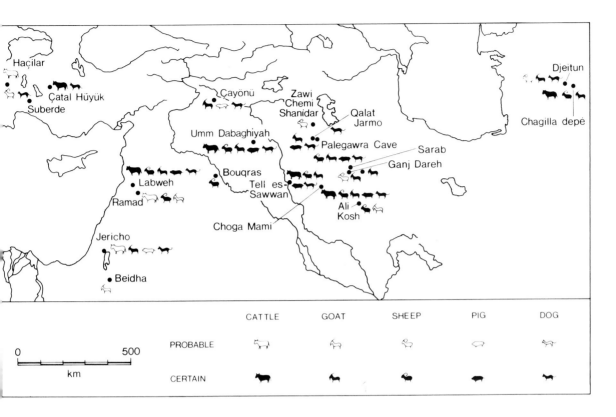

Figure 4.4   Archaeological sites in western Asia from which there is evidence for the earliest domestication of animals (after Bökönyi, 1976).

here, and in any case it is probable that plants were nurtured and animals kept in captivity all over south east Europe and western Asia long before their remains can be identified in the archaeological record. What can be said with certainty is that by the seventh millennium BC the domestic goat and sheep had become the principal source of meat and raw materials for the early farmers of this region, and that cattle and pigs were probably also undergoing the process of domestication at this time.

The ass and the horse were the last of the livestock animals to be domesticated and the history of their enfoldment into human society has been the most difficult to unravel. This must be in part due to the fact that the distribution of the wild progenitors, in North Africa for the ass, and central Asia for the horse, has not provided the ideal conditions for excavation and retrieval of animal remains that exist in the so-called 'fertile crescent' of western Asia where goat, sheep, pig, and cattle were all locally domesticated.

Short accounts of the present knowledge on the history of these livestock species, on which most of the peoples of the world have depended for their livelihood for at least the last six millennia, are given in the following chapters.

5 *Sheep and goats*

ORDER ARTIODACTYLA, FAMILY BOVIDAE, TRIBE CAPRINI

Wild caprines are very successful ruminants and in the absence of man they would probably inhabit the greater part of the mountains and hilly regions of Europe and Asia. North America too would have a high population of wild sheep but not goats which were never part of the indigenous fauna of the New World. The mountain sheep of North America and eastern Siberia differ from most of the races of wild sheep that inhabit western and central Asia in that they evolved, in the absence of goats, in adaptation to life on high mountain ranges and precipitous cliffs. Perhaps because the Asiatic sheep had to compete during their evolutionary period in the Pleistocene with the goat and ibex their natural habitat is at lower altitudes on open rolling mountains and foothill-slopes.

The mountain sheep of North America are rock jumpers like the ibex and there is no evidence that either of these caprines

Figure 5.1   European mouflon, *Ovis musimon*

(*Ovis canadensis, Capra ibex*) was ever domesticated. On the other hand the wild sheep and goats of the high lands of western Asia were probably the first livestock animals to be altered in their physical appearance by human control.

SHEEP

The taxonomy and nomenclature of the sheep and goats is very complicated, especially that of the sheep where altogether at least forty wild races have been described. There are, however, eight basic taxonomic groups of sheep that may be summarized as follows. They are grouped here according to their diploid chromosome numbers,* as in Nadler *et al.* (1973).

$2n = 52$	$2n = 54$	$2n = 56$	$2n = 58$
Ovis nivicola (Siberian snow sheep)	*Ovis aries* (domestic sheep)	*Ovis ammon* (arkhar-argali sheep)	*Ovis vignei* (urial)
	Ovis musimon (European mouflon)		
	Ovis orientalis (Asiatic mouflon)		
	Ovis dalli (dall or thin-horned sheep)		
	Ovis canadensis bighorn sheep)		

Ovis orientalis, the Asiatic mouflon, was most probably the ancestor of all domestic sheep and possibly also of the European mouflon, *Ovis musimon*. Both mouflons are dark-coloured sheep with rather small bodies, long legs, and relatively short tails. The horns are ringed rather than ridged and they are smaller and less twisted than in other groups of wild sheep. There is a black throat ruff, a small well-defined white rump patch and in some forms a well-marked saddle patch. The underparts are white. The Asiatic mouflon is redder in colour than the European; it is found in mountainous regions from Asia Minor to southern Iran.

Ovis musimon, the European mouflon, was until recently only found wild in the mountains of Corsica and Sardinia but it has now been successfully introduced as a wild animal to many European countries (Fig. 5.1). The origins of this sheep are not known. It was thought to be a truly wild species, a relic of the European wild sheep of the Pleistocene that survived only in the refuge area of the Mediterranean islands, but the lack of fossil evidence for sheep on the islands as well as in Europe weighs against this theory. It seems more probable that the mouflon, rather than being a relic of a wild species, is a relic of the first domestic sheep that were taken to Europe by the early Neolithic farmers sometime around the seventh millennium BC. In other countries these sheep were progressively changed by selective breeding but on Corsica and

*Chromosomes carry the genes that are the basis for heredity; they are microscopic thread-like bodies found in identical pairs (that is they are diploid) in the nuclei of the cells of plants and animals. There is usually a constant number of diploid chromosomes for each species.

Sardinia the sheep ran wild and have lived so ever since. Another better documented example of relic sheep of this type is the Soay sheep from the islands of St. Kilda in the Outer Hebrides of Scotland. These sheep are also very primitive but they exhibit characters in the fleece that prove them to be rather more domesticated than the mouflon. Soay sheep do, however, closely resemble the European mouflon which does have one character suggestive of man's influence; this being the rather high proportion of female sheep that are hornless. Hornless ewes are very common in domestic breeds of sheep and are also present in the Soay but they are very rare in all wild sheep, other than the European mouflon.

Figure 5.2   Urial, *Ovis vignei*

East of the range of the Asiatic mouflon the urial sheep, *Ovis vignei*, is found in the mountainous areas stretching from north eastern Iran to Afghanistan and north western India (Fig. 5.2). It is possible that present day domestic sheep may be in part descended from the urial but its higher chromosome number of 58 makes any direct ancestry unlikely. These are large sheep with horns that are ridged as well as ringed. The throat ruff is white and the tail patch poorly defined. In north central Iran *Ovis vignei* interbreeds with *Ovis orientalis* and there is here a hybrid race of sheep with mixed characters. Typically the horns of the urial are curved in one plane and point downwards below the neck, whilst in most Asiatic mouflons the horns are curved round and upwards above the neck.

The arkhar or argali sheep, *Ovis ammon*, are found in the mountains of central Asia and this group includes the magnificent, and now probably very rare, Marco Polo sheep (Fig. 5.3). The argali inhabits the Altai mountains and the Pamirs, Tien Shan, the Himalayas, and Tibet. They are adapted to life in higher mountains than the other more western groups of Asiatic sheep and they are very large animals with very widespread, massive horns which point outwards away from the body. It is unlikely that these sheep have contributed to the present day breeds of domestic sheep. It has recently been suggested, however, that modern stock could be improved and enlarged by interbreeding with argali sheep (Short, 1976).

Figure 5.3   Marco Polo sheep, *Ovis ammon poloi*

The wild sheep from eastern Siberia and North America fall into the final group although they have differing chromosome numbers (Fig. 5.4). That of the American sheep is the same as in domestic sheep ($2n = 54$) but it is most improbable that these species are in any other way connected with them. What is more likely is that the mouflons and the Siberian and American sheep represent two diverging arms of a geographical cline which is centred on the argali sheep of the southern Himalayas, the region where it is likely that evolution and expansion of sheep first began. This evolution has resulted in a reduction of the number of chromosomes and it is believed that the earliest forms of *Ovis* could have had 60 chromosomes as in the goat.

Figure 5.4   Bighorn sheep of North America, *Ovis canadensis*

Detailed behavioural studies have been carried out on the wild sheep of North America (*Ovis canadensis*, the bighorn) by Geist (1971, 1975) and on the urial sheep of the Himalaya by Schaller (1977), whilst in Britain a long term study of the feral sheep of St. Kilda (the Soay) is continuing (Jewell *et al.*, 1974). These studies have not shown up any great differences in behaviour between these separate groups of sheep except for the perhaps crucial point noted by Schaller (1977, p. 322) that the mountain sheep of America lack a submissive posture whilst Eurasian sheep have one. As has been said in Chapter 3 on domestication of the dog, it is the capacity for active submissive behaviour that enables personal relationships to develop between an animal and man.

Valerius Geist has said that, 'it is hard to imagine a wild animal more readily tamed than mountain sheep', but perhaps the wild mouflon living in western Asia 8000 years ago was indeed even easier to tame, and perhaps the wild goat was as well. Behavioural studies such as those carried out by Geist have shown that man can associate with wild sheep and goats in a way that is not possible with red deer or gazelle, to take two other species of ungulate that were of primary importance as providers of meat in the prehistoric period. The reasons for this are that the deer and gazelle are territorial animals which although they live in groups or herds do not have a social structure that is based on dominance hierarchies. Man, sheep, and goats, on the other hand, have a social system that is based on a single dominant leader, they have a home range but do not defend a territory in the same way that deer and most antelopes will. The difference between home range and territory in mammals has been explained by Jewell (1966) and its comprehension is essential to the understanding of the process of domestication of livestock animals.

Home range may be defined as a restricted area within which individual groups of animals live; it is the area, usually around a home site or core area, over which the animal normally travels in search of food. A migratory animal may have a winter home range and a summer home range and into this category herds of livestock animals such as reindeer fall, as do human pastoralists. The home range differs from a *territory* which is defended, usually by male animals. Many species of mammal, especially carnivores, will defend a territory within a home range. Sheep will maintain well-defined home ranges but they have no core area and do not defend a territory, whilst their social structure predisposes them to leadership by a herdsman. This enables them to be bunched up together in compact groups, often of both sexes, and they will even flourish better when crowded together.

Most species of deer, antelope, and gazelle do not behave like this which must explain why they were not domesticated in the ancient world, although some were known to be kept in captivity, as for example by the Ancient Egyptians. Many species are quite solitary in their behaviour or will be found only in small family groups, whilst others are gregarious or social but the males are territorial and would be difficult to manage under confinement. Furthermore these animals are adapted to escape from fleet-footed predators; they are nervous by temperament and being plains-living animals they can run extremely fast. They require sophisticated fencing to contain them and when constrained they will often panic and die of shock.

Sheep and goats would present none of these problems to the primitive farmer. Survival in the wild is for them a matter of finding enough food in a harsh mountainous environment rather than primarily a need to escape by swift flight from predators. They are therefore not so nervous, they cannot run so fast, and they can be easily constrained with a fence of thorn thickets or they can be trained to stay close to a farmstead or village of their own accord as their natural home range is relatively localized.

Ancient hunters undoubtedly followed herds of ungulates and observed their behaviour. It would be an advantage to keep the wild herds in a compact group to be preyed upon whenever meat was required and from this began the control and management of certain species which by their behaviour were 'predestined' for domestication. It is likely, therefore, that domestication of goat and sheep took place in two ways. One was by the herding and controlling of wild flocks that were perhaps kept in a favoured situation such as near a supply of water simply by habituation of a herdsman with a group of wild animals. The second method was by the taming and rearing of young animals that were imprinted on man as their leader.

The few sheep bones that have been identified from the pre-pottery Neolithic levels of Jericho (Clutton-Brock & Uerpmann, 1974) that are dated from 8000–7000 BC may well be the remains of animals that were controlled by man in either of these ways. Several other sites in western Asia have provided similar finds of sheep as has the early Neolithic site of Argissa-Magula in Greece which has a radiocarbon date of 7200 BC. As sheep were not indigenous wild animals in Greece in the post-Pleistocene these animals must have been taken there by man. The earliest dated site to produce remains of sheep that are assumed to have been under human control, if not actually domesticated is Zawi Chemi Shanidar in north east Iraq (see the map, Fig. 4.4). This site has a radiocarbon date of $10\,870 \pm 300$ BP (W-681) which makes it rather older than the Pre-pottery Neolithic A levels of Jericho from which two sheep bones have been retrieved. The remains of sheep from these early sites do not show morphological changes that indicate the influence of man. The supposition that they are from tamed or domesticated animals is therefore based either on their geographical context being wrong for the wild species or on the proportions of juvenile to adult bones being improbable for the relics of a hunting economy. Both these criteria are dubious and are open to other interpretations but on the other hand there can be little doubt that it was during the eighth and seventh millennia BC that man first began to domesticate sheep and goats within the region of western Asia and that these early pastoralists spread rather rapidly westwards into Europe and probably north and east into Asia and the Far East.

Morphological changes that can be seen in the sheep remains from later Neolithic sites (sixth and fifth millennia BC) are hornlessness in the females and a shortening of the limb bones. It is probable that the fleece changed from the wild type as a result of mutations followed by artificial selection, perhaps as early as the sixth millennium BC, for a statuette of a woolly sheep of this date has been found at Tepe Sarab in Iran.

The outer coat of wild sheep is stiff and hairy and covers a short woolly undercoat which only grows in the winter. The outer hairs are long and bristly and are known as kemps. In highly domesticated sheep these kemps are absent and the fleece consists entirely of the woolly undercoat which grows all the year round and is not shed in the summer. The wool may be brown, white, black, or a mixture of these colours. In wild sheep and in primitive domestic breeds the undercoat is shed each spring and it falls away in dense mats which can be gathered or plucked from the animal and made into felt or spun. The loss of this character of self-shedding is advantageous to the shepherd who needs the wool because it means none is lost while the animal is grazing, but on the other hand it does have to be manually sheared.

Most of the features of domestication in the sheep, these being alteration of horn shape (with hornlessness in the ewes), the fattiness and length of the tail, and the woolly, white fleece, were already common in western Asia by 3000 BC, for all these characters are shown in pictorial representation in Mesopotamia and are also written about in the Babylonian texts. In Ancient Egypt there were white, black and piebald sheep before 2000 BC.

GOATS

Whereas sheep are grazing ungulates that inhabit hilly regions and the foothills of mountains, goats are browsers whose natural habitat is on the high, bleak mountain ranges of Europe, Asia, and Ethiopia. They have never penetrated as far north as the mountain sheep which explains why goats did not cross the Bering Straits into North America when there was a land bridge between the two continents during the Pleistocene. Because of their adaptation to a particularly harsh environment goats are perhaps the most versatile of all ruminants in their feeding habits, a factor that has greatly affected their success as a domestic animal. They are also extremely hardy and will thrive and breed on the minimum of food and under extremes of temperature and humidity. The goat can provide both the primitive peasant farmer and the nomadic pastoralist with all his physical needs, clothing, meat, and milk as well as bone and sinew for artefacts, tallow for lighting, and dung for fuel and manure. Goats will complement a flock of sheep, which are perhaps usually rather easier to herd, by browsing on thorny scrubland whilst the sheep prefer the grass. Goats may have been of positive assistance to the Neolithic farmers in helping to clear land after the primary forest was burnt or cut down, and it is often argued that it has been the browsing of goats over the last five thousand years that has in great part caused the expansion of the desert areas of the Sahara and the Middle East.

Living goats can be placed in four groups according to the shape and curvature of their horns. The chromosome number is known for only two of these groups, the domestic goat, and the ibex, in both of which there is a diploid number of 60.

Capra hircus, the domestic goat. If horns are present they are either straight or twisted but they always have a more or less well-developed keel on the anterior edge (Fig. 5.5).

Capra aegagrus, the bezoar goat. The horn is curved like a scimitar and the anterior surface is compressed laterally so as to form a sharp anterior keel. As in all wild goats the horns of the male are very much larger than in the female, and the male has an inter-ramal beard (under its chin). The bezoar goat is found over much the same range as the Asiatic mouflon (*Ovis orientalis*), that is in the mountains of Asia Minor and across the Middle East to Sind. It is also found on some Aegean islands and in Crete, but again like the mouflon it is possible that these goats constitute relic populations

Figure 5.5   Feral goats browsing in woodland, Cornwall, England (photo J. B. & S. Bottomley).

Figure 5.6   Bezoar goat, *Capra aegagrus*

Figure 5.7   Markhor, *Capra falconeri*

Figure 5.8   Ibex, *Capra ibex*

Figure 5.9   Spanish ibex, *Capra pyrenaica*

Figure 5.10   East Caucasian tur, *Capra cylindricornis*

of very early domestic animals that were taken to the Mediterranean island during the prehistoric period (Fig. 5.6).

It is certain that this species of goat, if not the sole ancestor of the domestic stock was at least its main progenitor.

Capra falconeri, the markhor, has horns that are twisted either as a straight screw or in an open spiral. The back edge of the horn is keeled but the front edge is usually flattened at least at its most proximal part. The markhor is found on mountains from east Kashmir to the Hindu Kush and south to Quetta in Baluchistan. It is also found in the southern USSR, and may have been formerly much more widespread in the mountains of Asia. Some authors have postulated that this species is the progenitor of the domestic goats that have rather similar-shaped horns, as for example those that were common in Ancient Egypt and can be seen in parts of Africa at the present day. The lack of an anterior keel on the horn of the markhor may, however, belie any close relationship with domestic breeds (Fig. 5.7).

Capra ibex, the ibex, has untwisted scimitar-shaped horns as in the bezoar goat but they can be readily distinguished by the regular anterior ridges or bosses that mark the length of the ibex horn (Fig. 5.8). There are several subspecies of ibex, all of them living at very high altitudes, and all having these particularly ridged horns which are never found in the domestic goat. The ibex is the only wild goat to be found on the mainland of Europe. It inhabits the Alps and the mountains of Spain where it is classified by Corbet (1978) as *C. pyrenaica* (Fig. 5.9). In Asia the ibex is found from west of Lake Baikal to Turkestan and Kashmir. The subspecies *Capra ibex nubiana* lives on the mountains of Arabia, and another subspecies is found in Ethiopia where it is probably a relic of a much wider-spread group that ranged from Egypt and the Sudan.

Capra cylindricornis, the East Caucasian tur (Fig. 5.10), is a little-known goat that has horns that are almost round in cross-section and which are curved in a single open spiral. It is found in the eastern Caucasus mountains and the shape of its horns indicates that this species of goat is unlikely to have had any effect on the domestic stock. Another species of tur, *Capra caucasica*, with rather short widely-divergent horns, inhabits the western Caucasus.

Around 8000 BC the hunter-gatherers of western Asia began to change their way of life. They no longer exclusively hunted wild animals such as gazelle, wild cattle, pigs, and onagers; they began to keep goats. It was a change that may be taken as a turning point in the history of mankind and if for only this reason it must be justifiable to call the ensuing Neolithic period a 'revolution'.

Although there may be evidence from radiocarbon dating that the domestication of sheep preceded that of goats it is quite clear from the many sites from which their remains have been retrieved

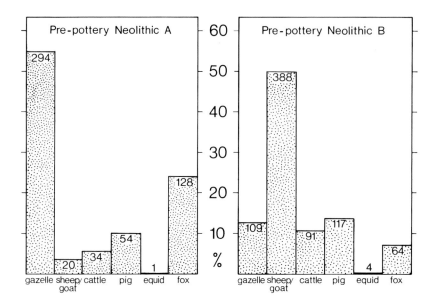

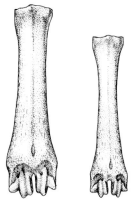

Figure 5.11   The numbers of animal remains from the Pre-pottery Neolithic A levels of Jericho (*c*. 8000 BC) compared with those from the Pre-pottery Neolithic B levels (*c*. 7000 BC). The numbers within the blocks refer to the absolute numbers of bones and teeth retrieved (see Clutton-Brock, 1979).

that the goat was more commonly kept as a supplier of meat at this earliest period than the sheep. Sites at which goat remains have been retrieved from the eighth millennium BC are shown on the map (Fig. 4.4). The change in the numbers of gazelle bones found compared with those of goat between the Pre-pottery Neolithic A period at Jericho and the Pre-pottery Neolithic B is shown in Figure 5.11. That the goats were under man's control whilst the gazelle were wild hunted animals is indicated by the morphological changes that can be seen in the bones and horn cores of the goats. No site has provided such evidence from the bones of gazelle.

The limb bones of goats from the Neolithic period (seventh and sixth millennia BC) are much reduced in size from those of the earlier period when the animals were either wild or just beginning to undergo control by man (Fig. 5.12). There is also a change in the shape of the horn core, especially in that of the male, which became more rounded and almond-shaped as well as smaller in diameter, and shorter. Hornlessness in goats is not nearly as common as in sheep and it is not known to have occurred until a much later period (hornless goats were present in the third millennium BC in Ancient Egypt, and were fairly common in Roman times).

The remains of ibex have been retrieved from only one Pre-pottery Neolithic site, this being Beidha in Jordan, south of Jericho (*c* 6800 BC), but even here there is no evidence that the ibex were domesticated for the horn cores were found in the shafts of graves together with a clay figurine of an ibex, and it could be that this goat was treated as a sacred animal rather than as a domestic one.

Little comparative work has been done on the social behaviour of the bezoar or scimitar horned wild goat, and the ibex, but it may be that the latter species being adapted to life as a rock jumper on

Figure 5.12   Metacarpal bones of goats to show the decrease in size from the wild to the domesticated form. Wild form 140 mm long, domesticated 100 mm.

Figure 5.14 In the earlier periods of Jericho goats with straight, scimitar horns predominated over those with twisted horns. In the later phases this situation was reversed, as shown in this diagram where the absolute numbers of horn cores are plotted on the vertical axis.

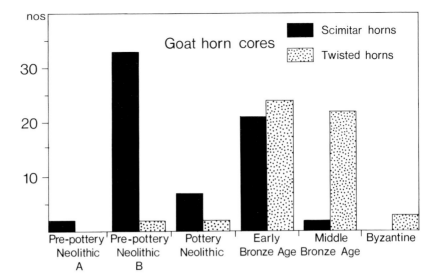

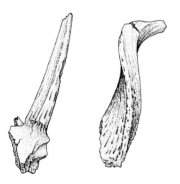

Figure 5.13 Straight and twisted horn cores of domesticated goats from Jericho. Straight core 150 mm long, twisted 180 mm.

high mountains would be less amenable to management and control by man than the bezoar. On the other hand the ibex certainly played a large part in the cultural life of the nomadic peoples of south Russia and central Asia for it holds a prominent position in their art, being particularly represented in the third millennium bronzes, and the Scythian gold treasure.

The earliest records of domestic goats all have straight scimitar horn cores but quite early on, twisted horns began to appear and by the Bronze Age in western Asia, twisted horned goats predominated over those with straight horns as they do at the present day (Fig. 5.13). The relative numbers of straight and twisted horn cores of goats from Jericho are shown in Figure 5.14. It is not known why goats with twisted horns should have been preferred to those with straight; it may simply have been because the farmers preferred them that way or it could be because billy goats can do less harm to each other when fighting if their horns are twisted, or it may be that there was a genetic link between the character for twisted horns and say a higher milk yield, or a more placid temperament. Such speculations, should not be given serious consideration without a background of information on behaviour, genetics, physiology, and anatomy, but they do highlight how much work there is still to be done on even the most common domestic animals.

6 Cattle

ORDER ARTIODACTYLA, FAMILY BOVIDAE, TRIBE BOVINI

To a modern west European farmer a cow is an animal that provides milk, butter, cream, cheese, and beef. To many peasant farmers throughout the rest of the world, today as in the past, a cow (or an ox) is a draught animal whose primary function in life is to draw a cart or a plough (Fig. 6.1). The cow may also provide a little milk and it will be killed and eaten when it becomes too old to work and breed any longer. Every part of the carcase is used, the meat and marrow for eating, the horns, bones, and hide for artefacts, weapons, and clothing, the fat for tallow (for lighting), the hooves for gelatine and glue; whilst from the living animal the manure is an essential part of the farming cycle and in some countries it is used as a fuel, and even as a building material. There are no other animals that provide such a versatile range of resources as domestic cattle and in addition both bull and cow have been symbolic figures in human societies for thousands of years.

The earliest firm evidence for the use of cattle as providers of milk comes from the ancient civilizations of Egypt and Mesopotamia and dates from the fourth millennium BC. It is probable that the practice of dairying spread from these centres, but it may also have arisen independently in northern Europe at an early period. There are, however, still today large areas of central Africa and eastern Asia where there is no tradition of milking and many adult people in these countries are physiologically unable to absorb lactose which is a main constituent of fresh milk (see the map, Fig. 6.2). In a recent review of this subject Simoons (1979) has suggested that the ability for adult humans to consume milk and dairy products evolved over a long period of time. Peoples who lived in countries where the economy was based on pastoralism but where they were subject to periods of nutritional stress would have a selective advantage if they were able to absorb and digest milk. Simoons has attempted to assess the prevalence of this ability in different countries of the world at the present day by tests on adult individuals for lactose malabsorption.

Traditionally, northern Europe is an area where dairying has played an important part in the economy and the numbers of

Figure 6.2 Areas of the Old World where there is no tradition of milking and lactose malabsorption in humans is common (from Simoons, 1979).

people found to have lactose malabsorption in these countries were very low. It is perhaps noteworthy, however, that the numbers rose markedly in southern Europe and were high in southern Italy and Greece. It may be remarked in this context that Columella and Varro, the Roman writers on agriculture made little mention of cattle in any other capacity than as draught or sacrificial animals; the Romans did not drink milk and this is reflected in the physiology of their descendants today.

Apart from those in south east Asia, domestic cattle are all descended from a single wild species, *Bos primigenius*, the extinct wild ox or aurochs (plural: aurochsen) (Fig. 6.3). The aurochs was a ubiquitous species that was very successful in the late Pleistocene and early Holocene and was widespread over most of the northern hemisphere with the exception of North America. The species declined in numbers as a result of hunting by man and was extinct in Britain, or at least very rare, by the Bronze Age. In central Europe it survived longer despite the onslaughts of hunters and it eventually died out in the early seventeenth century; the last individual is said to have been killed in Poland in 1627. The aurochs was a browsing and grazing ruminant that inhabited forests but could also flourish in open scrub. It was probably not found north of 60 degrees latitude or south of 30 degrees latitude except in India where a distinct form of wild ox,

Figure 6.1   Threshing on Chios, Greece with two cows and a mule (photo author).

Bos primigenius namadicus, has been recognized from the Pleistocene fossil record. It is possible that this form was the ancestor of the humped zebu cattle of India which later spread to Africa and Asia.

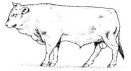

Figure 6.3   Reconstruction of *Bos primigenius* bull compared with its domestic descendant.

Such a wide ranging species as *Bos primigenius* is likely to have had a large number of geographical races that would have merged into each other as a cline but these are difficult if not impossible to recognize from fragmentary osteological remains. There is, however, some evidence for geographical variation in the coat colour as can be seen from the Palaeolithic rock paintings of different regions; those from Lascaux in the Dordogne, France, being particularly notable. From these Zeuner (1953, 1963) made the following deductions about the appearance and conformation of the northern and central European race: Height between 1·5 to 2·0 metres at the shoulder, the males with a very strong neck. The bulls were mostly black, becoming intensely black with age; the muzzle and some hair on the forehead may have been white and there was usually a white line running the length of the back. There was also sometimes a light saddle patch. The cows were more or less red in colour, as were the calves. The northern races were probably larger than those from more southern latitudes and they would have had heavier more woolly coats. Those from the south may have had longer limbs in relation to the size and weight of the body and some may have been quite pale in colour.

One problem that beset archaeologists and biologists trying to deduce the origins of domestic cattle in the early years of this century was that they did not realise that there was very marked sexual dimorphism in the aurochs and they therefore interpreted the remains of bulls and cows as coming from two different species. The bulls were very much larger than the cows with longer, differently-shaped horns.

Although *Bos primigenius* can be accepted as the single progenitor of the domestic cow it is only one of several species of bovid that may be classified as cattle. These are grouped in the tribe Bovini (family Bovidae) and a number of other species within the tribe have also been domesticated. Some of these are not well known in the west but they are of great economic importance in eastern Asia. The domestic and wild groups are given here with their conventional Latin, binomial names in order to keep confusion to the minimum. There is controversy amongst taxonomists about the generic status of some members of the Bovini because it has been shown that interbreeding can occur not only between species but also between some genera. Here, however, for the purposes of clarification it is easier to retain the more traditional nomenclature (see also Appendix I).

EUROPEAN CATTLE

Bos primigenius primigenius. The aurochs or extinct giant ox.

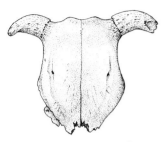

Figure 6.4   This skull of a small bull exemplifies the type described by Owen (1846, p. 508, Fig. 211) as *Bos longifrons*. Both this skull and Owen's specimen were retrieved from peat bogs in Ireland. They are nowadays assumed to be the remains of prehistoric domestic cattle descended from the aurochs, *Bos primigenius*. Skull width 190 mm.

The wild progenitor of *Bos taurus*.

Bos taurus. Domestic humpless cattle.

[*Bos longifrons* (synonym: *Bos brachyceros*). Not a discrete, recognizable form. Described on the remains of small, short horned cattle from archaeological sites in Europe. These remains were originally believed to be from a separate wild species that was presumed to be the ancestor of domestic cattle. This hypothesis was based on the theory that *Bos primigenius* was too large and ferocious an animal (it was described as such, for example, by Caesar *c.* 54 BC) to have ever succumbed to man's domination. As, however, remains of the so-called *Bos longifrons* have never been found in deposits earlier than the Neolithic it may be safely assumed that they represent domestic animals (Fig. 6.4).]

ASIATIC CATTLE

Bos primigenius namadicus. The extinct Indian form of the aurochs. Sometimes treated as a separate species, *Bos namadicus*. Probably the ancestor of the Indian humped cattle or zebu, *Bos indicus*. Represented by a few fossil skulls found on the Indian sub-continent.

Bos indicus. Domestic humped cattle, the zebu and its derivatives.

Bos gaurus. The wild gaur found in the forests of India and south east Asia. Sometimes classified in the subgenus *Bibos*. Probably the progenitor of the mithan, *Bos frontalis*.

Bos frontalis. The domesticated mithan or gayal. Sometimes classified in the subgenus *Bibos*. Found in Assam, north west Burma and Nepal.

Bos javanicus (synonym *Bos banteng*). Sometimes classified in the subgenus *Bibos*. The wild banteng is now very rare and is restricted to a few isolated populations on Java. Previously it was found on Borneo and in Malaysia and Burma. Domestic banteng, known as Bali cattle, are found on Bali, Timor, Celebes, Java, and Borneo.

Bos sauveli. Sometimes classified in the subgenus *Bibos*. The kouprey, a little-known wild bovid from the forests of Cambodia. Now an endangered species.

Bos mutus. The wild yak which is now an endangered species and very rare. It is found in Tibet, Nepal, and the Himalayas. Sometimes classified in the subgenus *Poëphagus*. Progenitor of the domestic yak, *Bos grunniens*.

Bos grunniens. Domestic yak used in the same mountainous regions where the wild species previously ranged.

BUFFALOES

Bubalus arnee. The wild Indian water buffalo is a rare animal today, found only in Nepal and south east Asia. It was formerly widespread and may have reached into western Asia. Progenitor of the domestic water buffalo, *Bubalus bubalis*.

Bubalus bubalis. Domestic water buffalo, the most common form of domestic cattle in many Asian countries.

Anoa species. There are three species of dwarf buffalo that are sometimes classified as subgenera of *Bubalus.* They are rare wild animals found on the Celebes and on Mindoro Island in the Philippines.

Syncerus caffer. The wild African buffalo which has never been domesticated. Found throughout the savanna lands and forests of the African continent.

BISON

Bison bonasus. The wild bison or wisent of Europe, at the present day restricted to a few forests in Poland and Russia. The bison was sometimes referred to in the old literature as 'urus' or 'auroch' and it is necessary when reading early accounts of wild cattle in Europe to ascertain which species is meant, *Bison bonasus* or *Bos primigenius.*

The European bison will interbreed freely with the American bison and will produce fertile offspring. It is therefore becoming accepted to treat the two forms as conspecific and they are then both named *Bison bison.*

Bison bison. The American bison, often called 'buffalo' in the United States. Formerly widespread over the whole continent but today only surviving in protected herds (see Chapter 17). Neither the European bison nor the American bison has ever been domesticated.

The origins of the domestication of the true cattle (*Bos taurus* and *Bos indicus*) are discussed in the remaining part of this chapter whilst other species of domesticated cattle are described in Chapter 14.

As with goat and sheep the remains of the earliest domestic cattle have been found on archaeological sites in western Asia and south eastern Europe. The sites are shown on the map (Fig. 4.4). It is probable, however, that cattle were not fully under human control until perhaps a millennium after the domestication of goat and sheep. There is no evidence, from decrease in size of the bones from the Pre-pottery Neolithic levels of Jericho for domestication or even of taming of the *Bos primigenius* that was probably quite common in the region. The earliest certain evidence for domestication comes from the site of Çatal Hüyük in Turkey, dated to *c.* 6400 BC, although further west the earlier site of Haçilar has also provided bovine remains that are rather smaller than the usual *Bos primigenius.*

It is rather difficult to imagine how or why the change from hunting of wild cattle to husbanding of tame ones began. A large bovid, standing at least 1·5 m at the shoulder would not make an easy captive, nor would the animal be easy to restrain unless it wished to remain close to human habitation. Again, it is difficult to surmise what the humans could use the captive bovids for in the early stages of domestication. It is very unlikely that a tamed aurochs cow would allow itself to be milked, because considerable effort and guile has to be put into persuading a cow of an unimproved or primitive breed to let down her milk. The ways in which this is done in various parts of the world have been reviewed, together with historical evidence for the same methods, by Amoroso & Jewell (1963). The cow must be quite relaxed and totally familiar with the milker, her calf must be present, or a substitute that she identifies with the calf, and it is often necessary to stimulate the genital area before the milk-ejection reflex will allow secretion.

Perhaps the earliest subjugation of cattle occurred by the encouragement of free-ranging animals to remain near a human settlement. This could be done by keeping supplies of salt and water available at fixed places with which the animals could become familiar. Simoons (1968) has described how mithans in the Assam hills can be persuaded to return from the forests and remain near the villages by exploiting their craving for salt. The wild free-ranging animals can be approached and even stroked whilst they are fed by hand with salt. These cattle are not milked or eaten and are kept only for purposes of sacrifice and barter and it is possible that the aurochs was originally tamed for the same functions. Many early prehistoric societies undoubtedly venerated cattle and for example at Çatal Hüyük there is positive evidence for the ritual treatment of cattle in the form of horn cores associated with human figurines and fertility symbols. One shrine from level VI (*c.* 5950 BC) had a bench in it with horn cores of *Bos primigenius* set in clay (Fig. 6.5).

Figure 6.5   Shrine VI.61 from Çatal Hüyük VI (from Mellaart, 1975).

A group of cattle encouraged to remain in the vicinity of a human settlement would affect the habitat in a number of ways. Firstly, being browsing and grazing animals, they would destroy the lower branches of trees, and bushes and they would trample the undergrowth and foul any natural source of water such as a pool or stream. Trampling of the land could be a positive advantage to the human inhabitants in that it would help to open up the landscape and enlarge the man-made clearings.

Secondly the cattle would totally destroy any crops sown or waiting to be harvested by the villagers unless these were securely enclosed or fenced in. Thirdly the presence of cattle and calves would attract predators such as the large cats, wolves, and bears.

Wild adult *Bos primigenius* would not be very vulnerable to attack by wolves or even by lions as, like the African buffalo today, they would be protected by the cohesive, aggressive behaviour of the herd. On the other hand a few individual cows, perhaps with calves, used to being fed by hand and probably not in peak condition, would fall easy prey, and it is hardly necessary to reiterate that no human community likes to have lions or wolves attracted to the outskirts of its settlement.

So the Neolithic farmers who were becoming increasingly dependent on cattle and who were perhaps beginning to use them as draught animals found that at the same time they were quite troublesome to keep. During the day the cattle had to be kept away from the water supplies and from growing crops, whilst at night they had to be protected from predators. The only way this could be done was by driving the cattle into a fenced compound, or kraal as is done throughout Africa today, or by keeping them in a specially-built cattle house or shed which was often built onto the farmer's house. Irrespective of the method of enclosure the animals were unable to feed at night and the loss of night-feeding has a very deleterious effect on ruminating animals such as cattle that are physiologically adapted to periods of grazing followed by periods of rumination. In addition the smallest animals would be the easiest to handle and the easiest to house. It is therefore not surprising that a dramatic decrease in the size of cattle occurred during the Neolithic period when they were first domesticated, both in western Asia and in Europe. This decrease in size was progressive until, during the Iron Age, cattle were bred that would be considered to be dwarfs by today's improved standards. Their withers' height was little more than a metre, although the oxen (castrates) that were used as draught animals would have been somewhat taller. This is because when a male animal is castrated the bones will continue to grow in length for a longer period than is usual for the entire male, so that oxen are taller but more fine-limbed than bulls and their horns are also longer and more cow-like in shape. In northern Europe these diminutive cattle were probably mostly black in colour and probably resembled quite closely the modern Dexter breed.

In the southern parts of the old world, in Egypt and Meso-potamia, where agriculture was more highly advanced than in the north, distinctive breeds of cattle were differentiated by the Bronze Age, the lyre-horned cattle of Ancient Egypt being the most obvious example (Fig. 6.6). Herodotus writing in about 450 BC described polled (naturally hornless) cattle bred by the

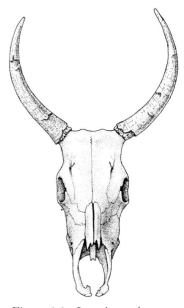

Figure 6.6   Lyre-horned cattle: skull, still with the horn sheaths intact, from Abydos, VIth Dynasty (*c.* 2181–2050 BC), see also Fig. 2.1.

Scythians, and from North Africa he described a breed of cattle with long, low-sweeping horns that must have closely resembled the modern British Longhorn. He describes* these cattle as follows:

In the Garamantian country are found oxen which, as they graze, walk backwards. This they do because their horns curve outwards in front of their heads, so that it is not possible for them when grazing to move forwards, since in that case their horns would become fixed in the ground. (IV, 183)

As with other livestock the Romans improved their breeds of cattle and were responsible for importing new stock into northern Europe. Columella had this to say† about the different kinds of oxen that he knew of:

... cattle show variation in bodily form and disposition and the colour of their hair according to the nature of the district and climate in which they live. Those of Asia and of Gaul and of Epirus are different in form, and not only are there diversities in the various provinces, but Italy itself shows varieties in its different parts. Campania generally produces small, white oxen, which are, however, well suited for their work and for the cultivation of their native soil. Umbria breeds huge white oxen, but it also produces red oxen, esteemed not less for their spirit than for their bodily strength. Etruria and Latium breed oxen which are thick-set but powerful as workers. The oxen bred in the Apennines are very tough and able to endure every kind of hardship but not comely to look upon. (VI, 1)

Figure 6.7   Zebu cattle

This description of local breeds by Columella is especially interesting in that it shows how the intermingling of artificial and natural selection produced different kinds of cattle that were best adapted for their particular environments and for the ploughing of different types of soils. It also indicates how ancient the present day remnants of local breeds are likely to be.

Humped or zebu cattle (*Bos indicus*) form a separate group from the European breeds (Fig. 6.7). They are morphologically distinct in having a longer narrower skull, a heavy dewlap, long legs, long pendulous ears, and a muscular or musculo-fatty hump carried over the back of the neck, or withers. They are often light in colour and are physiologically better adapted to a tropical environment than are humpless cattle. There is no general agreement on whether zebu cattle were first developed in south western Asia or on the peninsula of India but there is little doubt that the many breeds of humped cattle in Africa at the present day are of secondary origin and were first introduced from India or the Middle East.

Humped cattle are depicted on cylinder seals from the ancient civilizations of Mohenjo-Daro and Harappa in the Indus Valley which are dated 2500–1500 BC, whilst in southern Iraq on Sumerian and Babylonian sites they are also depicted from about

*Translated by Rawlinson (1964, I, p. 358).
†Translated by Forster & Heffner (1968, p. 125).

the same period. The neural spines of the posterior thoracic vertebrae of the zebu (lying behind the region of the hump) are bifurcated (Fig. 6.8) and if present in a collection of animal remains from an archaeological site this bifurcation allows the assumption that humped cattle are represented, although the bifurcated spine can occasionally be found in European breeds of cattle. Fragments of bifurcated spines, the oldest of which dates to 1400 BC have recently been described from the Tell of Deir'Alla in Jordan and it can be reasonably assumed that these came from zebu cattle (Clason, 1978).

Although all the different kinds of humped cattle can be conveniently grouped under the one name 'zebu' there are as many breeds with humps as there are without, and it is probable that they are equally as ancient in origin as the European breeds.

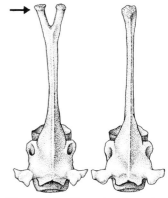

Figure 6.8   Posterior thoracic vertebrae of a European and a zebu ox to show the bifurcation in the neural spine of the zebu, arrowed.

7 Pigs

Figure 7.1   Wild boar, *Sus scrofa* (photo Geoffrey Kinns).

Figure 7.2   Distribution of the subspecies of *Sus scrofa*.

Sus scrofa

1	*castilianus*	15	*leucomystax*
2	*barbarus*	16	*moupinensis*
3	*meridionalis*	17	*chirodontus*
4	*majori*	18	*riukiuanus*
5	*reiseri*	19	*taivanus*
6	*scrofa*	20	*cristatus*
7	*falzfeini*	21	*andamanensis*
8	*attila*	22	*nicobaricus*
9	*libycus*	23	*jubatus*
10	*sennaarensis*[1]	24	*vittatus*
11	*nigripes*	25	*floresignus*
12	*raddeanus*	26	*timorensis*
13	*ussuricus*	27	*papuensis*[1]
14	*coreanus*		

[1] Probably reverted to a wild state

After several authors

ORDER ARTIODACTYLA, FAMILY SUIDAE

Domestic pigs are all descended from one species, the wild boar, *Sus scrofa*, still a relatively common wild animal found in many countries throughout Europe, Asia, and North Africa, but not in North America (Fig. 7.1). In Britain the wild boar has been extinct since at least the seventeenth century AD. Although it is such a ubiquitous species the boar flourishes best in broad-leaved deciduous forest areas so that as a wild animal it must have been at the northern edge of its range in Britain. Extermination of the wild boar in Britain, as in other countries, has occurred, however, as a result of overhunting by man and artificial deforestation rather than from any deterioration in the climate that may have occurred since the early Holocene.

About 25 subspecies of *Sus scrofa* have been described, these representing varieties that have evolved in adaptation to local conditions of environment and climate (see the map, Fig. 7.2).

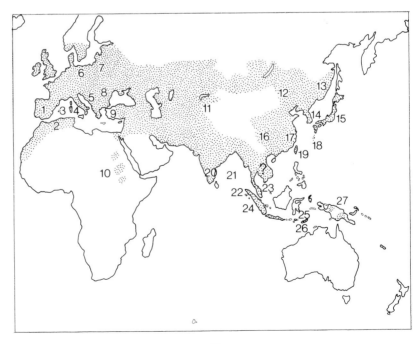

An osteological character that has been much used for the distinction of the different subspecies of wild pig is the relative length and shape of the lacrimal bone (Fig. 7.3). It has often been stated that the south east Asian pigs belonging to the *Sus scrofa cristatus* and *Sus scrofa vittatus* groups have a shorter lacrimal bone, whilst the subspecies of the European wild boar have a long lacrimal bone as can be seen from the figure. This bone, being from a rather solid part of the skull, is often found amongst the food remains from archaeological sites and it has therefore been tempting to use this convenient character as a basis for deducing the origins of various groups of prehistoric domestic pigs; some being said to be derived from the Asian subspecies and some from the European. For example Zeuner (1963) following the classic work of Rütimeyer in the mid-nineteenth century believed that the small domestic pigs from the Neolithic Lake dwellings in Switzerland (the so-called Turbary pigs) had a separate origin from another group of larger, locally domesticated pigs because they had short lacrimal bones. As the first change to occur in the domestication of the pig is a shortening of the frontal region of the skull and a general diminution of the size of the animal it is much more likely that a shortening of the lacrimal bone is correlated with this process rather than that it reflects the locality of the wild progenitor.

The remains of pig from the Pre-pottery Neolithic B levels of Jericho (*c.* 7000 BC) are marginally smaller than those of the preceding A levels and so it could be postulated that domestication or at least human interference with a wild population of pigs was beginning at this time (Clutton-Brock, 1979). Similar evidence has been found at Jarmo and at the site of Argissa-Magula in Greece and it is probable that the earliest settled 'farmers' began to manage wild pigs at the same time as they husbanded the first sheep and goats. But the pig was not a very highly favoured food animal in comparison to the goat at this early period in eastern Europe and western Asia. In northern and western Europe, however, at this time (culturally the Mesolithic) wild pig together with red deer provided the primary source of meat. Two to three thousand years later as the early farmers spread from the east into Europe they brought the domestic pig with them and at first pigs and cattle were more successful and more common than domestic sheep and goats. This would follow because the landscape was still heavily wooded and the environment was better suited to pigs and cattle than to caprines. Later, towards the end of the Neolithic and the beginning of the Bronze Age, sheep began to take over as the most common species of domestic livestock.

During the early postglacial period in northern Europe the wild boar must have been an exceedingly common animal perfectly adapted to the vast areas of deciduous forest and liking especially the river valleys and marshy places. It is easy to

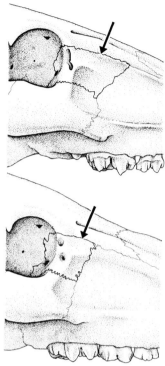

Figure 7.3   The lacrimal bone (arrowed), in *Sus scrofa scrofa* above, and *Sus scrofa cristatus*

picture how the early Neolithic peoples would interbreed their valuable imported stock with tamed local animals for although adult wild boar can be ferocious the piglets are very easily tamed and take readily to life in association with a human settlement.

Pigs are much more like dogs and people than they are like cattle or sheep and in many respects their behaviour seems to be intermediate between that of the carnivores and the more highly evolved artiodactyls. They are very versatile in their feeding habits and will survive well by scavenging on most of the foods that dogs and humans can live on. Also like carnivores but unlike all other artiodactyls, except the hippopotamus (to whom they are fairly closely related), pigs seek out and enjoy bodily contact with other members of their family groups; they like to huddle together (Fig. 7.4, p. 78). This may be related to their method of reproduction for, as is well known, pigs have large litters of up to ten young at a time, whilst all other artiodactyls will usually have only one young, or at the most three, at a time. Another character that the pig shares with carnivores is the habit of nest building and bed making. All pigs will root out moulds or wallows for themselves and the females will build a nest in which to give birth to their young. First the sow will dig out a mould the size of her body, then she will collect twigs and grasses, carrying them in her mouth, and will drop them into the hollow. She will carry material a considerable distance and will move grass and leaves from the edges by digging them in with her feet. When the nest reaches a certain height the sow will cover it with branches that may be up to two metres in length and she will root about and move round and round in a circle until the nest consists of a comfortably lined round or oval heap. She enters this, lies down and gives birth on her side, another difference from other artiodactyls in which parturition usually occurs whilst the female is standing.

Within minutes or even seconds of birth young wild piglets will try to reach their mother's teats and during the first day they will establish a nipple order, that is each piglet will establish itself on a nipple and will always feed from this one and will defend it against other piglets. This behaviour is not often seen in domestic pigs, but explains why it is sometimes difficult to introduce piglets from another litter to a foster mother.

Another way in which pigs resemble carnivores rather than artiodactyls is their weak physical development at birth. Most ungulates are born with well-coordinated sense organs and musculature so that they can stand and even run a few hours after birth, but a litter of piglets is much more immature and has to stay within or near the protection of the nest for the first few weeks. When they do leave the nest the sow keeps in contact with the piglets by nose-to-nose touching at frequent intervals and a complicated series of vocalizations. If a piglet is lost all the others will squeal and try to find it, whilst the mother will become excited and when she finds it will touch the piglet with her snout. Pigs do not, however, lick their young in the way that carnivores do (Frädrich, 1974).

Feeding and sleeping are the two activities that dominate the life of the pig, but whereas ruminants will feed for a short time, then ruminate, then perhaps sleep for a while and then feed again, pigs will feed continuously for many hours and then sleep for many hours. This means that in captivity they do not require food during the night, and their sleeping and feeding programmes can be easily accommodated to that of humans. Despite their liking for close bodily contact, wild pigs do not in nature congregate in large herds in the way that cattle and caprines do; the sows and their young stay together as a family group, whilst the boars are solitary except at the rutting time. The number of sows that a boar will keep together for mating depends on how many he can defend against other males, usually in present day groups of wild pig it is less than ten. In past times, however, when domestic pigs were put out to pannage and allowed to roam free-range in the European woods, a herd of one hundred or more sows with a few boars was not uncommon.

Male pigs can be very aggressive and they will fight to defend their sows or they will fight over food (rather as dogs will) but like other species of mammal that have been domesticated *Sus scrofa* is not territorial which means that they will succumb to interference by man. Since Roman times, to prevent fighting, it has been a common practice to castrate young domestic boars and also to spay the females that are required for fattening but not for breeding. Columella* describes the operation of removing the ovaries but he obviously did not approve of it.

Domesticated pigs can be easily trained to come where they are wanted by means of feeding them and Varro† describes how in Roman times they were trained to follow the sound of a horn:

> The swineherd should accustom them [piglets] to do everything to the sound of the horn. At first they are penned in; and then, when the horn sounds, the sty is opened so that they can come out into the place where barley is spread out. . . . The idea in having them gather at the sound of the horn is that they may not become lost when scattered in wooded country.
>
> (II, iv, 17–20)

Since ancient times there have been two methods of keeping domestic pigs, firstly the system of allowing large herds of animals to roam at will in the forests under the eye of a swineherd who was often a child armed only with a stick, and secondly there was, and still is today, the house or sty pig (Fig. 7.5). The forest pigs were, in the Roman period, and also for at least a millennium before and after this time, dark skinned, long legged pigs that were very small, perhaps half the size of the wild boar. They bred quickly and their large numbers made up for their small size which was adapted for survival in an open wooded environ-

---

* Translated by Forster and Heffner (1968, p. 291).
† Translated by Hooper and Ash (1967, p. 363–4).

Figure 7.5 Forest pigs and sty pig below

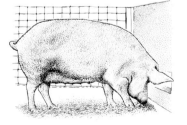

ment where there was much competition for the natural foods that were available all the year round. Columella* writes thus about the ideal land on which to keep these pigs:

The most convenient feeding-grounds are woods covered with oaks, cork-oaks, beeches, Turkey oaks, holm-oaks, wild olive trees, terebinth-trees, hazels, junipers, nettle-trees, vine-tendrils, cornel-trees, strawberry-trees, plum-trees, Christ's thorn, and wild pear trees.
(VII, ix, 3–7)

It is no wonder that the Romans had a saying that, 'the race of pigs is expressly given by nature to set forth a banquet'; salted hams made from such well-fed pigs must indeed have been good food.

The sty pig was a bigger animal and even in the Roman period it could be smooth, white, and so fat that it could hardly stand. The pig has always been the poor man's animal and in post-medieval Britain fat bacon was the most provident and common source of meat. Almost every family kept a sty pig that was fattened on household scraps and then re-cycled as ham, pork, bacon and dripping.

Towards the end of the eighteenth century AD a transformation occurred in the European breeds of domestic swine following the introduction of small, quick-maturing, light-boned pigs from China and south East Asia. These were crossed with the native breeds and the resulting improvement in the stock was so great and the swine so popular that within fifty years there were probably few pigs left in western Europe that were not at least

* Translated by Forster and Heffner (1968, p. 293).

Figure 7.6 Fat-bellied pig, Bali (photo I. Glover).

partly descended from Asiatic imports. It is therefore unlikely that there are any breeds remaining that are pure-bred descendants of pigs domesticated in western Europe during the prehistoric period.

In the Far East the pig has been of even greater economic importance than in the West and it is probable that in China the oldest domesticated animals were the pig and the dog, both being reared as providers of meat. The extreme fat-bellied pigs of the East are of very ancient origin and many varieties are bred in China where Epstein (1969) has stated that there are over one hundred breeds of pig at the present day (Fig. 7.6). To the Pacific islanders the dog and the pig have been the chief suppliers of meat for many thousands of years and on some islands the culture is closely interwoven with rituals concerning these two species of domestic animals. Titcomb (1969), amongst other authors, has described how the women of Papua, New Guinea, until recently, would suckle puppies and piglets along with their own infants as a matter of course.

There is evidence to suggest that the pig may have been domesticated in the Far East as early as the present archaeozoological records indicate its inception in western Asia. Wild pig was not part of the indigenous fauna of Papua, New Guinea and yet on an archaeological site excavated by Professor J. Golson there have been found hollows in an ancient land surface dated to 9000 years ago that have every appearance of being 'fossilized' pig wallows or moulds.

Nobody knows why one species of animal is favoured as a supplier of meat to one nation, whilst to another it is considered a taboo animal, unclean and untouchable. Such preferences and taboos can last for thousands of years; to the Pacific islanders the pig and the dog were the mainstay of their lives, whilst to the Muslims and Hebrews they are forbidden food, and in Europe the dog has not been eaten during the Christian era except under conditions of extreme famine.

It may be that the taboos against the pig in western Asia originated in Ancient Egypt where as described by Herodotus:*

> The pig is regarded among them as an unclean animal, so much so that if a man in passing accidentally touch [*sic*] a pig, he instantly hurries to the river, and plunges in with all his clothes on. Hence, too, the swineherds, not withstanding that they are of pure Egyptian blood, are forbidden to enter into any of the temples, which are open to all other Egyptians; and further, no one will give his daughter in marriage to a swineherd, or take a wife from among them, so that the swineherds are forced to intermarry among themselves.   (II, 47)

Although the pig was considered an unclean animal in Ancient Egypt, perhaps as early as the third dynasty (*c.* 2686 BC), pork was allowed to be eaten on certain days and the animals were sacrificed to Bacchus and the moon (according to Herodotus) so it is not obvious how strict the taboos really were.

The animal remains from the excavations of the Sumerian sites of Tell Asmar and Abu Salabikh in northern and southern Iraq show that at the beginning of the third millennium BC (somewhat later than the Third Dynasty in Egypt), small, almost dwarf, pigs were common domestic animals and were a primary source of meat. There is no satisfactory explanation at present as to why they were later so abhorred but we have here clear evidence of how the pig, more than any other animal, has been subjected to the whims of human taste and religious scruple.

* Translated by Rawlinson (1964, 1, p. 137–8).

Figure 7.4   Wild boar sow suckling her piglets, see p. 73 (photo Geoffrey Kinns).

Figure 8.2   'Reconstituted tarpan, *Equus ferus gmelini*', Poland, see p. 81 (photo G. Barker).

Figure 8.8   Scythian horsemen, apparently riding with stirrups, portrayed on the finials of a gold torque, fourth century BC Crimea, excavated 1830. Hermitage, KO 17, see p. 89 (photo Lee Bolton).

Figure 9.1 Domestic
donkeys, Iraq, see p. 91
(photo author).

Figure 9.2 Semite with his
donkey. Beni Hasan, Tomb 3,
c. 1900 BC, see p. 91 (from
Davies & Gardiner, 1936,
photo BM).

Figure 10.6 From 'fowling
in the marshes'. A cat
amongst birds, Thebes, 1420–
1411 BC, see p. 111 (from
Davies & Gardiner, 1936,
photo BM).

8 Horses

The horse was the last of the five most common livestock animals to be domesticated, and as a species it has been the least affected by human manipulation and artificial selection. This is because the Equidae are probably genetically less variable than the pigs, cattle, and caprines, but also because the horse has only one important use to man, that of carrying him and his belongings from place to place in the shortest possible time. Therefore there has been little or no selection for the animal resources necessary to human economies such as superfluous fat, increased milk supply or the development of the trunk of the animal for meat at the expense of the limbs. The horse evolved as a fast-moving ungulate capable of migrating over great distances and it is its inherent potential for speed and strength that has been exploited and developed by man.

When horses were first domesticated during the late Neolithic period they were probably husbanded as a food animal, for their chopped bones are found on archaeological sites along with the remains of other kitchen debris, but it cannot have been long before the early farmers found that horses could be loaded with goods to be carried, ridden on, and trained to pull carts. It is probable that cattle were used before horses for ploughing and probably also for draught, but once the horse became established as a means of transport there was a transformation in the way human beings lived. This happened, it seems, surprisingly late, during the second millennium BC in Europe and only a millennium earlier in southern Russia and western Asia. The world was opened up to the horse-rider; he could travel anywhere and with the aid of the ensuing improvements in the techniques of warfare he could conquer the new lands into which he moved. On the other hand to train a group of men, let alone a cavalry, to ride even the diminutive horses of the ancient world required the coordination of technological skills that took time to develop.

The history of the wild horse in Europe and Asia from the end of the Pleistocene until its domestication in perhaps the fourth millennium BC is poorly understood. During the late Pleistocene large herds of wild horse, *Equus ferus*, must have been common on the open plains of Europe, Asia, and North America and they were extensively hunted by Palaeolithic man. At the end of the Ice Age their range was very much reduced probably as a result of the spread of forests combined with human predation. In America the horse may have been extinct 8000 years ago although some people believe it lingered on until the post-Columbian period. In Europe the wild herds were gradually pushed eastwards into the semi-deserts of central Asia until today they are represented only by *Equus ferus przewalskii*, the wild horse of the Mongolian steppes (Fig. 8.1).

It could well be that the horse was a species of equid that evolved during the Pleistocene and at the end of that geological period was doomed to extinction by changing climatic and ecological conditions. By chance it was saved by man who re-introduced it into its former habitat under his protection and as a tamed animal. Now it is not only the domestic horse that has been reintroduced, for during this century a number of wild Przewalski horses have been brought from Mongolia and now breed freely in zoos and parks. It is probable, however, that if it had not been domesticated the species *Equus ferus* could have been extinct by the second millennium BC. This fate did overcome another species of equid that inhabited southern Europe and western Asia during this period. This equid, named *Equus hydruntinus*, is known only from the fossil record and so its outward appearance can only be a matter for speculation, but its bones and teeth indicate that it was about the size of a wild ass and it was more like an ass or a zebra than a true horse.

Another subspecies of *Equus ferus* that was not as fortunate as Przewalski's horse, in that it is now extinct, lived wild on the steppes of the Ukraine until the end of the last century. This was the tarpan, *Equus ferus gmelini*, which was said to be wild in western Europe in the middle ages and was exterminated in Poland in the eighteenth century AD. There has been some controversy, however, about the tarpan as some writers believe that it was not a true subspecies of wild horse but represented a population of feral ponies. Its appearance is known from pictures and from a skull and skeleton in the Soviet Academy of Sciences in Leningrad. Attempts have been made to 'reconstruct' the tarpan by cross-breeding from primitive native ponies, and several herds of these 'new tarpans' now flourish in Poland (Fig. 8.2, p. 78).

Although it looks as though *Equus ferus* would not have survived long into the post-Pleistocene period there are of course other species of equid that are more viable, the most successful of which are the zebras of Africa, followed by the wild asses of Asia, and the African ass (progenitor of the donkey) which is now near extinction. Although the zebras and asses will also be exterminated by man unless there are strong government policies to protect them these equids are not in danger from natural changes in the environment as occurred in temperate Europe and

Figure 8.1   Przewalski horse, *Equus ferus przewalskii* (photo Geoffrey Kinns).

Figure 8.3   Exmoor pony
(photo S. Gates).

North America in the early Holocene. All species within the family Equidae are grazing herbivores and the true horse is an inhabitant of temperate, well-watered grasslands, so that if driven away from this preferred habitat its survival becomes tenuous.

Although Zeuner (1963) was convinced that the tarpan was a truly wild horse descended from Pleistocene ancestors, and that it was this equid that was the main progenitor of European domestic horses, this seems unlikely to other writers, including Kowalski (1967), because they cannot envisage how the tarpan could have survived during the prehistoric period as a wild species in the dense forests of central Europe. The same argument may be applied to the Exmoor ponies of Britain which are often claimed to be directly descended from British Pleistocene stock (Fig. 8.3). It seems more probable that these ponies, and perhaps also the tarpan, constitute populations of feral animals, perhaps of very ancient origin.

Because of the wide variation in size and conformation of modern domestic horses, with small stocky ponies in northern Europe, heavy horses in north and central Europe and slender-limbed Arabian horses in the south, it has often been suggested that at the end of the Pleistocene there were several different species of wild equids that were domesticated. For example, Sanson as early as 1869, believed that there were eight types of horse, each with a different ancestor. Later than this, Ewart (1907) classified horses into three types: firstly the *Steppe* variety that he allied with Przewalski's horse and which was said to be long-headed; secondly the *Forest* variety comprising ponies with short, broad heads, and thirdly the *Plateau* variety that contained the horses with long, narrow heads. This category, rather oddly, combined the 'Celtic' ponies (including the Exmoor pony) from a subarctic environment, and the 'Libyan' from a subtropical climate.

Other writers have divided the three basic categories of domestic horse into 'cold-blooded' which are the heavy draught horses and 'hot-blooded' which are the Arabian horses, whilst the 'warm-blooded' are the cross-breds like the British *Thoroughbred*. This may be a useful way of classifying breeds of horses for the owner of modern stock but in biological terms all that it means is that horses, like most other mammals, follow Allen's Law which summarizes how animals are affected by climate and how they adapt to it. In the northern latitudes mammals tend to be large and heavy-bodied with short legs relative to the size of the body and they have small compact extremities such as the ears. In the lower latitudes and hotter climates mammals have longer legs, longer ears, and finer-limbed proportions. They also have a shorter, sleeker coat all the year round whilst northern mammals will grow a thick hairy coat in the winter, as do the so-called 'cold-blooded' horses. Despite artificial selection for characters quite other than those required for adaptation to special environments it can be seen that many breeds of domestic livestock conform to this pattern. To see this we need only compare the long-legged, lop-eared, thin skinned sheep of the Middle East with say a British long-woolled breed such as the Lincoln which is quite barrel-shaped. Yet both these breeds have been developed from the same progenitor.

The closest relatives of the Equidae are the rhinoceroses and tapirs and in evolutionary terms these three families form a more primitive group than the ruminant artiodactyls. The equids are grazers and their high-crowned teeth and digestive tract are specialized for the assimilation of grasses, a food material that relatively few mammals can live on exclusively. The asses of Africa and Asia will also browse but the true horse is a specialized inhabitant of grasslands and it is poorly adapted to a desert or forest biotope although it can survive if forced to take refuge in these environments. In Mongolia where Przewalski's horse inhabits the harsh, cold desert-steppe this horse must be on the extreme edge of its natural range. As a result it is an unnaturally small horse with a deep mandible and cheek teeth that are very large for the size of the skull (Fig. 8.4). This could indicate that this subspecies is derived from an ancestral Pleistocene form that was considerably larger, for it is usual for reduction in the size of the body to precede reduction in the size of the teeth. This is because the teeth have less genetic variability and so 'lag behind' other parts of the body in morphological and size change.

It is unlikely that Przewalski's horse is directly linked to the ancestors of the European domestic horses. It is more likely that it is a side shoot from the main line of Pleistocene horses and that it survived extinction because, in the early Holocene,

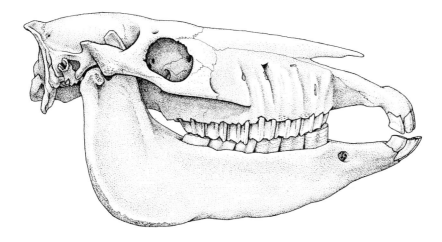

Mongolia provided a retreat from the encroaching forests and human hunters. Similarly the tarpan, if it was a truly wild horse, was the survivor of a post-Pleistocene evolutionary line. We can therefore only guess where the progenitor of the domestic horse came from and what it looked like. What does now seem clear is that there were very few populations of wild horses inhabiting western and northern Europe after the end of the last glaciation (*c*. 8000 BC). In Hungary there have been no finds of horse from archaeological sites until the end of the Neolithic (Bökönyi, 1974). Around 2000 BC, however, horse remains are found throughout Europe from the Orkney Islands to Greece, and they greatly increase in numbers throughout the ensuing Bronze Age.

It is probable that the wild stock from which all these domestic animals were bred inhabited the plains of southern Russia, from the Ukraine to the region of Turkestan. The earliest domesticated horses spread out from this arc, probably first of all being traded as a food animal but rather quickly supplanting the slower ox as a draught animal. It has been suggested that the people who first tamed horses in the area north of Iran were influenced by the Middle Eastern civilizations which already used tamed onagers, *Equus hemionus*, for draught but this is a controversial point for which there is little evidence on either side.

The first domestic horses may have resembled the tarpan in appearance, that is they were the size of large ponies, of medium build, and with upstanding manes. From this stock all the different types and breeds of horse that are known today were developed as a result of artificial selection in combination with natural selection for adaptation to local climatic and environmental conditions. So that, for example, mountain-living ponies have small neat hooves that are sure-footed on narrow stony paths, whilst plains-living horses are larger and have wider hooves, and the ponies from marshlands, such as the Camargue in the south of France, have extraordinarily wide-splayed hooves.

Figure 8.4   Skull of Przewalski horse. Skull length 470 mm

There is no need to assume separate ancestry for these breeds of horse any more than there is for the Persian fat-tailed sheep and the British longwool.

On the other hand, it is possible that there was a geographical cline in the populations of wild horses with those in the northern part of the range being smaller and more sturdy than those in the south. The reason for believing this is that even in the earliest finds of domestic horse there are considerable differences in the size and proportions of the bones from different regions. Horse remains from Neolithic and Bronze Age sites in Britain, for example, are from animals with the proportions of an Exmoor pony, whilst those from the two sites in Ancient Egypt that have yielded the earliest equid remains (Buhen and Thebes, *c.* 1500 BC) are from larger more slender-limbed horses (Clutton-Brock, 1974).

Early domestic horses show less change in size and general morphology from the wild animals than do other species of livestock and there appears to have been no dramatic decrease in size at the onset of domestication, as can be seen for example between *Bos primigenius* and domestic cattle. The archaeological sites in Russia that have produced large numbers of remains of horses provide little evidence to distinguish domestic from wild animals. It is therefore rather surprising that by 1500 BC, an early phase in the history of man's domination of the horse, there seems already to be definition into northern pony and Arabian types. These differences, however, were certainly not so marked as at the present day. Most domestic horses, throughout their range in the ancient world, were less than 145 cm in withers height (14 hands) and the great majority, even during the classical period, judging from depictions on pottery and friezes, were less than 125 cm in withers height.

Until recently little attempt has been made to study the behaviour of horses and asses in the wild and even less attention has been paid to the comparative behaviour of domestic equids. In a short review Klingel (1974) states that two types of social organization have been observed in the Equidae. Burchell's zebra, *Equus burchelli*, the mountain zebra, *Equus zebra*, the horse, *Equus ferus przewalskii*, and possibly the Asiatic wild ass, *Equus hemionus*, do not establish territories. These equids live in family groups of a stallion with as many mares as he can muster and their young. Stallions without mares live in separate groups. They will not try to entice away adult mares from the family group but will abduct the juvenile mares to form their own families when possible.

The second behavioural group is territorial and consists of Grevy's zebra, *Equus grevyi*, and the African wild ass, *Equus africanus*, and in these species the sexes separate for part of the year. This zebra and the ass are both inhabitants of desert and desert-steppe where water is usually restricted and grazing sparse so that the holding of territories is likely to be an adaptation to this specialized environment. The territories are extremely large and are defended by the males only when there is a female in oestrus near the boundary. In these equids there are no permanent bonds between the adult animals; they are found living solitary or in a variety of different associations that can change from day to day.

In the non-territorial equids (which includes the horse) there are well established dominance hierarchies within the family. The stallion may occasionally lead the group, especially during migration, but it is usually the most high-ranking mare that is the leader and the others will follow in order of their dominance, with the stallion bringing up the rear. The foals, often more than one for each mare, will follow their mother in order of their birth with the youngest first. This adherence to an order of dominance whilst moving from place to place explains how a convoy of domestic horses or a cavalry charge can be held together with very little effort by the human rider or driver who assumes the position of the stallion. It also explains why it is necessary to castrate all domestic stallions except those required for breeding, unless they are stabled on their own and kept away from other horses. A free-living stallion will try to assert his dominance over other animals in a herd and will try to abduct young mares from other herds. Mohr (1971) wrote that she observed no differences between the behaviour of domestic horses and wild Przewalski horses except that stabling of the stallions inhibits their natural behaviour. A wild stallion will bite and kick his mares in order to make them follow his will and it requires no mean skill on the part of man to assert his dominance over such an animal and to break it in for riding. In general it may be stated that the domestication of wild cattle, *Bos primigenius*, which are naturally highly social animals would have been a great deal simpler than the taming of wild horses which have a highly developed system of dominance hierarchies but are not inherently so gregarious.

It is noteworthy that in Ancient Egyptian and classical paintings where stallions are depicted harnessed to chariots the necks of the horses are shown strongly arched and the head is held high by means of a bearing rein. Mohr (1971, p. 75) puts forward the theory that this was not so much for the sake of appearance and style but in order to prevent the stallion lowering his head into the threat and attack position (Fig. 8.5).

Differentiation of domestic horses probably occurred in the Bronze Age and by the Iron Age two distinct groups were well established in the archaeological and early historical records. One group consisted of the small 'Celtic' ponies of Britain, western Europe and Greece, whilst the second group consisted of larger horses from Scythia and the Russian steppes. It is not true that the 'Celtic' pony was an especially developed breed or that the

Figure 8.5   Horses held with a bearing rein: 'Head Dress of a riding horse' (above) 'The king in his chariot returning from battle' (opposite) (Both from Layard (1849 vol. II).)

Figure 8.6   Miniature Arab pony on the Royal seal of Darius, probably from Thebes, c. 500 BC (photo BM).

Celtic people had ponies that differed from others in Europe, but certainly horses were of importance in the Celtic economy and they figured largely in the household and religious art of the Iron Age. These ponies were very small, some being less than a metre high at the withers.

It is from this period that the earliest written accounts of riding and the management of horses in Europe survive, including the works of Xenophon who lived from 430–354 BC. Philip of Macedon (the father of Alexander the Great) followed the advice of Xenophon in the equipping and training of his cavalry and even today there is much that can be learned from this author and soldier's *Treatise on Horsemanship*. Within recent years a wealth of literature has been published on the history of the wheel and on the horse in peace and war. To repeat it here would be out of place but there are, however, a few points that may be made in relation to the two groups of horses that are now known to have inhabited Europe and Asia in the first millennium BC.

In the early civilizations of Egypt and Mesopotamia the chariot was the only approved means of transport for the elite, that is the hunters, warriors and kings. The chariot was common too in ancient Greece and throughout Europe and the horses that were harnessed to chariots belonged to the small group as can be seen from numerous pictorial representations. In the west of Europe they were small stocky ponies whilst in western Asia they had the appearance of miniature Arabs (Fig. 8.6).

To the north east of Greece, in the ancient world, lay the kingdom of Scythia (see the map, Fig. 8.7) where nomadic herdsmen lived who were a source of wonder to the Greeks and whose way of life was described in some detail by Herodotus*.

* Translated by Rawlinson (1964, 1, p. 294–9).

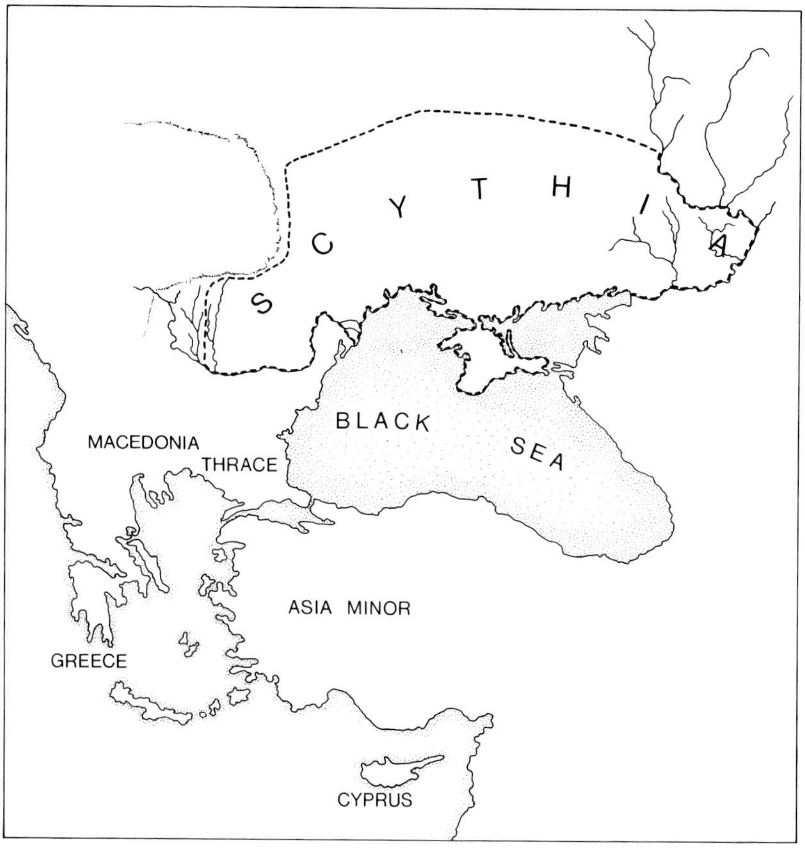

Figure 8.7   The Ancient World

The Scythians were non-literate but they had a highly sophisticated form of art based on the stylization of animal motifs which included countless representations of their horses. One of the most realistic of these is the frieze on a vase from the fourth century BC kurgan of Chertomlyk in the Ukraine (quoted from Bökönyi, 1968). The frieze shows horses being handled and broken in by their Scythian riders and it can be seen that there are marked differences between these animals and the western 'Celtic' ponies. The Scythian horses, as is known also from their mummified remains found in the famous frozen tombs of Pazyryk (Rudenko, 1970), were characteristically large and Arabian in their conformation.

At this period Scythian horses were imported into Greece in great numbers, often with their riders, as recorded for example in the seizing of 20 000 mares by Philip of Macedon, and it is probable that the large Roman military horses were bred from stock that came originally from Scythia, perhaps via Greece. During the Roman period the remains of large horses, some more than 145 cm high at the withers, are found on many archaeological sites all over the Roman empire and as far north as Hadrian's wall in Britain.

Alexander the Great learned his horsemanship from the Scythians but it is perhaps a rather surprising fact that neither he nor Julius Caesar knew of the stirrup. Maybe it was for this reason that the cavalry was not very important to the Romans, but three hundred years earlier Alexander conquered two million square miles of the ancient world with his Companion cavalry-men. Before the wide-scale use of the stirrup battles had to be fought at close range because only the most highly skilled horseman could shoot a long distance projectile from the back of a galloping horse without being thrown off himself. The cavalry surged into the enemy and attacked them with javelins, swords, and daggers and Alexander's successes were based on the formation of a wedge-shaped cavalry charge that pierced the centre of a pitched battle (Lane Fox, 1973).

It is generally believed that stirrups were first used by the Chinese and they are first mentioned in the literature in AD 477. After this time the stirrup spread slowly west in the early post-Roman period but did not reach western Europe until the eighth century AD. It may be, however, that the Scythians sometimes rode with stirrups at a much earlier period for the ends of a very fine gold torque cast in the classical style are in the form of two horsemen whose feet are clearly placed in stirrups which appear to have consisted of a metal hook attached to the saddle by a metal chain (Fig. 8.8, p. 78). Even if the stirrup was known to these highly skilled nomadic horsemen, whose lives were centred on their animals, it did not spread to Greece, for it is never seen

Figure 8.9 Alexander the Great riding his horse Bucephalus and wearing the lion's head helmet of Heracles attacks Persian horsemen (photo Hirmer Fotoarchiv).

in any pictorial representation and Xenophon would certainly have written about stirrups if he had ever seen them.

Instead Xenophon* gives the following advice on the correct seat for riding:

> When he has taken his seat, whether on the horse's bare back or on a cloth, we do not like that he should sit as if he were on a carriage seat, but as if he were standing upright with his legs somewhat apart; for thus he will cling more firmly to the horse with his thighs, and, keeping himself erect, he will be able to throw a javelin or to strike a blow on horseback, if it be necessary, with greater force.   (7,5).

Later on in this work Xenophon describes how weapons should be launched from horseback and it can be seen from the frieze on the 'Alexander sarcophagus' exactly how this was done in battle (Fig. 8.9). The sculpture shows Alexander on his horse Bucephalus attacking Persian horsemen at the battle of Issus. Xenophon wrote (translator's spelling):

> To inflict injury on an enemy we recommend the short curved sword rather than the long straight one; for from a horseman, seated aloft, a blow from a scymitar will be more effective than one from a straight sword. Instead of a read-like spear, as it is weak and inconvenient to carry, we rather approve of two javelins of corneil wood; for a skilful thrower may hurl one of these, and use the other against assailants either in front or flank, or rear . . . We shall intimate in a few words how the javelin may be hurled with the greatest effect. If the rider advance his left side, at the same time drawing back his right, and rising on his thighs, and launch his weapon with its point directed a little upwards, he will send it with the truest aim, if the point, as it is discharged, is directed steadily to the mark.   (12, 12–14).

The immigrant nomads from the north who made up the greater part of Alexander's cavalry were indeed centaurs. How else could a man ride a galloping horse, manage all the accoutrements of primitive metal armour, carry spare weapons and hurl a javelin, all without stirrups?

---

* Translated by Watson (1884).

9 *Asses, mules, and hinnies*

ORDER PERISSODACTYLA, FAMILY EQUIDAE

With the possible exception of the cat, the continent of Africa has produced only one domestic mammal – the donkey (Fig. 9.1, p. 79). Archaeological records of the remains of the wild African ass from the early Holocene are scant and there is no material evidence for when asses were first tamed. This is because, as with the horse, there was probably no discernible difference in the size of the bones of the wild ass and the early domestic form. It is likely, however, that by the fourth millennium BC the ass was being bred in captivity by the Ancient Egyptians, although there are very few early sites from which ass bones have been retrieved. These are not recorded here because the finds were poorly documented and their dates are not established. From the time of the fifth dynasty (2500–2345 BC) the domestic ass was frequently depicted in Egyptian art, as is shown on the wall painting from the famous tomb of Beni Hasan (Fig. 9.2, p. 79). The donkeys were often shown carrying loads or with saddle cloths on their backs.

Within modern times the wild ass, *Equus africanus* (see Appendix I for nomenclature), has been restricted in its distribution to North Africa, but it is possible that in the late Pleistocene and early Holocene the ass also inhabited Arabia and the drier parts of the Levant. The wild ass may indeed have evolved in Asia and subsequently moved into Africa, or like the hartebeest, *Alcelaphus buselaphus*, which had a similar distribution, it may have spread into western Asia from Africa. Remains of the hartebeest have been found on a number of sites in the Levant ranging in date from the Upper Pleistocene to the Bronze Age when it became restricted to Africa (Fig. 9.3). The remains of the hartebeest in western Asia are easily identified but those of the true ass are much more problematical because Asia is the homeland of another group of asses, *Equus hemionus*, the onagers or hemiones (six geographical races or subspecies of which are known, as well as the more distinct kiang), and the bones and teeth of the two species are not always easy to distinguish.

In addition another equid, *Equus hydruntinus*, inhabited

Figure 9.3 Hartebeest, *Alcelaphus buselaphus*

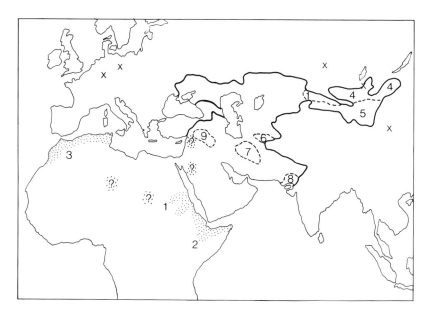

Figure 9.4   Distributions of wild asses, *Equus africanus* and *Equus hemionus*, as far as they are known within the Recent period.

southern Europe and western Asia until the late prehistoric period. Little is known of this extinct equid although its bones and teeth are fairly often found on archaeological sites. It was not a horse but whether it lay nearest the asses, the hemiones, or the zebras has not yet been established and the few representations from Palaeolithic rock art that are suggestive of an ass-like animal do not help with its specific relationships.

Taxonomists have classified the asses and hemiones under a variety of genera, species, and subspecies, some combining them together and some elevating the asses to the separate genus *Asinus* whilst retaining the hemiones in the genus *Equus* with the horses. For purposes of clarity the more traditional classification is used here of including all the equids within one genus, *Equus*. (This includes the zebras but these are not listed as they have never been domesticated). Short summaries are given below of the species and subspecies of wild asses and hemiones that are living today or have become extinct since the end of the Pleistocene; their distributions are given in the map (Fig. 9.4):

EQUUS HYDRUNTINUS
An extinct ass-like animal of which little is known. It was hunted by man up to the Neolithic but as far as can be ascertained it was never domesticated.

THE AFRICAN ASSES
Equus asinus, the domestic ass or donkey.
Equus africanus '*atlanticus*', the Algerian wild ass (the subspecific name is invalid as it was earlier used for a zebra). This ass was formerly an inhabitant of Algeria and the Atlas Mountains. It has been identified as being represented on Roman mosaics from North Africa and was probably exterminated by the Romans.

Whether this race was imported into southern Europe and bred with domestic stock is not known but it is likely to have been. The mosaics depict this ass with a strongly marked long shoulder stripe and with bars on the legs.

Equus africanus africanus, the Nubian wild ass, is believed now to be extinct but was previously distributed over the mountainous semi-deserts of Nubia and the eastern Sudan from the Nile to the shores of the Red Sea. The last individuals seen in the wild by Europeans were shot in the 1920s and 1930s but several Nubian asses survived in zoos until more recently. The Nubian ass, shown in Figure 9.5, was reddish-grey in colour, had a well-marked dark stripe running along the length of the back, and a sharply-defined, but very short, dark stripe across the shoulders. There were no horizontal bands around the legs.

Figure 9.5   Nubian wild ass, *Equus africanus africanus*

Equus africanus dianae, the Eritrean wild ass, was probably inter-mediate between the Nubian and Somali asses. It had a long but poorly-marked shoulder stripe, and faint traces of bands around the legs.

Equus africanus somaliensis, the Somali wild ass, is the only race to survive at the present day, but it too is near extinction. This ass is larger than the Nubian subspecies and may have a shoulder height of over 140 cm. It is buffy-grey in colour in the summer turning to iron grey in the winter and has a contrasting light belly and legs. The stripes along the back and shoulders are ill-defined, although quite commonly there is a long thin shoulder stripe. On the other hand there are well-marked horizontal bars around the lower limbs, as in Figure 9.6. Today this race is only found in northern Somalia and there may be a few in the Danakil region of Ethiopia.

Figure 9.6   Somali wild ass, *Equus africanus somaliensis*

Another subspecies of wild ass was described by the explorer Heuglin in 1861 from the Red Sea coast, and was named by him *Asinus taeniopus*. It is doubtful, however, whether this ass was really a true race; it is more likely to have been a population of cross-bred feral animals, for herds of feral donkeys are common all over the habitable parts of the Sahara and are often confused with the wild form. They are, however, smaller and are much more mixed in colour.

In ancient times it may be assumed that the distribution of the geographical races of the wild ass was contiguous and that they formed a cline running across North Africa from the Atlas Mountains to the Red Sea. It is not possible to say which of the known subspecies contributed most to the stock of present-day domesticated donkeys; the Nubian ass may have been domesti-cated by the Ancient Egyptians, whilst it is probable that the Romans imported the Algerian ass.

THE ASIATIC ASSES

Equus hemionus hemippus, the Syrian onager, is now extinct. It was the smallest of the hemiones and it inhabited the alluvial

plains of the Near East from the Levant to Iraq. The subspecies
is known from the descriptions of hunters and travellers, from
a few skins and skeletons in museum collections, and from the
archaeological record both pictorial and osteological. The Syrian
onager represented the most western subspecies of a geographical
cline that until recent times ranged from the Levant across south
western Asia to Nepal and north to Chinese Turkestan and
Mongolia.

The Asiatic asses are usually redder in colour than the African
and although they may often have a dark line running along the
back they only occasionally have a shoulder stripe. The limbs are
rather more slender for their length than those of the African ass
and the ears are shorter (although they are still longer than in the
horse). The tail is tufted as in the African ass and donkey, and
the mane is short and erect.

Equus hemionus onager, the Persian onager, is a larger subspecies
than the Syrian (Fig. 9.7). Within recent years it has become
an endangered race and is found in small numbers in north
eastern Iran. Formerly the Persian onager was widespread
throughout Iran at a higher altitude than the Syrian onager. It
has been much hunted from the prehistoric period until the
present. Because of its preference for a higher altitude than the
Syrian onager, it has been suggested that it was this race that
inhabited Anatolia until the classical period and was described
by Pliny. Its remains have been found on a number of archaeological sites in the Zagros mountains, in Iran.

Figure 9.7　Persian onager,
Equus hemionus onager

There are four other subspecies of hemione that inhabit Asia
east of Iran. These are, *E. h. hemionus* from northern Mongolia,
E. h. luteus from the Gobi desert, *E. h. kulan* from Turkmenia,
and *E. h. khur* from the Thar desert. In addition there is another
hemione which is nowadays usually separated as a different
species, *Equus kiang*. The kiang inhabits the Tibetan plateau; it is
the largest of the wild asses and like the yak it is adapted to life
at very high altitudes.

HYBRIDS

All the species of equid will interbreed, including the zebras;
they will produce viable offspring but these are very rarely fertile,
with the exception of those from a wild African ass, *E. africanus*,
and a domestic donkey, *E. asinus*, which are fully fertile. Similarly
a domestic horse will produce fertile progeny from crossing
with a wild Przewalski horse, although their chromosome numbers
are different. The diploid chromosome numbers (see footnote
p. 53) of the Equidae are as follows :-

$$2n = 66$$
Equus ferus przewalskii

$$2n = 62$$
Equus africanus

$$2n = 64$$
Equus caballus
(domestic horse)

$$2n = 62$$
Equus asinus

$$2n = 56$$
Equus hemionus

$$2n = 46 - 32$$
Zebras

Hybrids produced by the artificially induced interbreeding of different species of equids may be described as truly man-made animals, and it is perhaps as the father of the mule that the donkey has made its greatest contribution to human economies. The mule is a perfect example of hybrid vigour – as a beast of burden it has more stamina and endurance, can carry heavier loads, and is more sure-footed than either the ass or the horse.

Different species of Equidae will not normally interbreed in nature and it requires the guile and expertise of man to bring it about. A *mule* is the offspring of a male donkey (jackass) with a female horse (mare), whilst a *hinny* is the offspring of a female donkey (jenny) and a male horse (stallion).

It is possible that in the ancient civilizations of western Asia both the donkey and the horse were interbred with the onager, *Equus hemionus*. There are no common names for the offspring of these crosses so they are described here simply as ass/onager and horse/onager hybrids.

The hybrids can of course be either male or female and the sexual organs, both internal and external, are fully developed although the animals are almost always sterile. In the ancient world, as at the present day, the production of mules was easier to manage and more successful than the production of hinnies.

Mules and hinnies have certain consistent characters as shown in Figure 9.8.

In general resemblance to the parents it may be said that the head and front end of the hybrids are like the sire, whilst the hindquarters are like the dam. The enamel patterns on the biting surfaces of the teeth are a combination of ass and horse in their characteristics. The body size is larger than the ass and may also be larger than the horse. Mules therefore have a heavy head with large ears and neat front limbs, whilst the hind limbs may be stronger and the tail is either covered with long hairs coming from the root as in the horse or at least has long hair for most of its length.

Figure 9.8   Mule, above, and hinny

The hinny may be distinguished from the mule by its lighter head and shorter, more horse-like ears, whilst the tail is tufted only at its end as is that of the donkey. These characters can be seen from representations in ancient art as well as in living animals. In the lower picture of the tomb painting of equids from Thebes in Egypt (eighteenth dynasty, *c.* 1400 BC) there is a very accurate depiction of what I believe to be a pair of hinnies

(Fig. 9.9). These have previously been erroneously identified as onagers because of their light coloration, small ears, and tufted tails. It could be that the artist's intention was to paint the Nubian ass but under no circumstances could these animals be onagers because of the painting of a strongly-marked shoulder stripe that is on the body and is not part of the harness. The upper picture is of a pair of stallions and is typical of the Ancient Egyptian horses of this period with strongly arched neck, short mane, and flowing tail. It should be noted of course that there is no band painted across the shoulders of this horse.

This is as far as I know the only ancient picture of a hybrid from Egypt but mules were very often represented in Mesopotamian art of the first millennium BC and are especially clear on the Assyrian reliefs from Ashurbanipal's palace at Nineveh (now in the British Museum). These reliefs are of hunting and battle scenes and are dated to about 645 BC. They have been most beautifully and accurately carved and provide a manifold source of information on the animal world as it was known to the Assyrians. Figure 9.10 depicts a very healthy-looking male mule.

Figure 9.9   Tomb paintings from Thebes *c.* 1400 BC. The top painting shows horses drawing a chariot. The equids in the lower painting are more difficult to identify but they could be hinnies (photo BM).

THE MULE IN THE ROMAN WORLD

Mules were an essential part of life to the Romans. They were used for riding, by farmers for ploughing and for drawing carts, and by the army for carrying baggage; their remains have been found on archaeological sites in many parts of Europe and recently a lower jaw of a mule has been identified from a Roman deposit in London (Armitage pers. comm.).

Figure 9.10 A mule laden with hunting gear from the palace of Ashurbanipal, Nineveh, *c.* 645 BC. Now in the British Museum (photo BM).

Columella[*] gave precise instructions on how a jackass that was to be used for fathering mules had to be taken from its dam and put with a suckling mare as soon as it was born. The donkey foal had to be reared with horses so that its behaviour patterns would become adapted to theirs and it would respond to a mare that was in oestrus. In order that a jackass could copulate with a mare that would be a taller animal than itself, the Romans built a special 'machine'. This consisted of a sloping wooden cage on to which the mare was harnessed. 'A badly-bred ordinary donkey' would first be brought to the mare to judge her compliance and only when it was ascertained that she was ready for mating would the chosen sire be substituted:

for unless she has already had experience of a male, she repulses the donkey with her hooves when he leaps upon her, and the affront which he has received inspires him furthermore with an aversion for all other mares. (VI, xxxvii, 9–11)

When the hybrid foal was born it was left with its mother for a year and then it was

put to feed far away in the mountains or in wild places, so that it may harden its hoofs and presently be fit for long journeys.

THE ASS AND ONAGER IN PRE-ROMAN WESTERN ASIA
As well as the interbreeding of the ass with the horse that has been carried out since ancient times, and is still common in some

[*] Translated by Forster and Heffner (1968, p. 221).

Figure 9.11 'Standard of Ur', battle scene (photo BM).

countries today, there is historical and pictorial evidence to indicate that the ass was also crossed with the onager, *Equus hemionus*, in western Asia in Sumerian and later periods (Zarins, 1976). As already described the mule and the hinny bear intermediate characters between the ass and the horse that are consistent and easily recognized, but the hybrids between the ass and the onager, because they are anyway close to each other in appearance, are more difficult to identify. It is for this reason that ass/onager hybrids have not until recently been recognized in Mesopotamian archaeology.

The wild horse, *Equus ferus*, did not inhabit the Near and Middle East after the end of the last Ice Age except in the most northern part of the region and except perhaps as an occasional immigrant southwards. Occasional finds of the remains of true horse in Anatolia and in the Middle East from the prehistoric period have posed a controversy, not yet resolved, as to whether they come from wild or early domesticated horses. It was the onager, *Equus hemionus*, that was the indigenous wild equid of western Asia and for the last thirty years it has been assumed by archaeologists that it was this equid that was domesticated by the ancient civilizations of the Fertile Crescent. Some authors have gone so far as to suggest that domestication of the horse only followed after the peoples of the north had gained experience of using onagers for riding and as draught animals.

There is one rather large flaw in this supposition. If the onager had been domesticated why is it not the common domestic equid of Asia today? We know that it is not, because, apart from its appearance, the domestic ass of Asia will not produce fertile offspring when crossed with an onager whereas it will when crossed with a European donkey or African wild ass. Therefore the donkeys of Asia have originated from *Equus africanus* and not from *Equus hemionus*. There is only one large mammal that we know was domesticated in the ancient world and is no longer associated with man, this being the North African elephant which was exterminated by the Romans after being tamed and exploited by them for a relatively short period.

The answer is that it is most unlikely that the onager was ever a fully domesticated and artificially selected animal. This contention is supported by the comprehensive review by Zarins (1976) who has examined every aspect of the history of equids in Mesopotamia during the third millennium BC. The onager was not domesticated because its inherent behaviour patterns could not be manipulated by man to produce a docile, submissive animal that would breed readily in captivity over many generations. It will be shown in later chapters how some groups of ungulates, notably the deer and gazelles, have not been domesticated because they are too 'nervous' to be closely controlled. The same holds true for the onager; all writers since Roman times have told of its bad temper and irascible nature, and yet an 'ass' is the most commonly represented equid in Mesopotamian art of the third and second millennia BC. Furthermore this 'ass' sometimes bears a closer similarity to *Equus hemionus* than it does to the true ass, *Equus asinus*. Figures 9.11–12 are typical examples; they are from the 'Standard of Ur', a hollow box covered with mosaic decoration that was excavated by Sir Leonard Woolley from the royal cemetery of Ur in southern Babylonia. The object was made in about 2500 BC; it represents a battle and a victory scene, and there are seven separate pairs of equids depicted on the sides. Those shown in Fig. 9.11 appear to be rather more like onagers than asses; the head is ass-like but the ears are rather short, the legs are long and slender, and the tail, although tufted, has rather more hair than is usual for the donkey. These could be ass/onager hybrids. On the other hand the two pairs of equids in the victory scene are more like donkeys because they have a clearly marked shoulder stripe (Fig. 9.12).

Figure 9.12   'Standard of Ur', victory scene (photo BM).

Every archaeologist has to be able to speculate and to build models from the fragmented evidence that comes from the soil. For equids in ancient Mesopotamia this evidence indicates the presence of an ass-like animal that was known to everyone. Therefore it was reasonable to assume that the local wild onager, a common enough animal, was domesticated by the Sumerians and it was this equid that was represented in the ancient art and in the accounts written in cuneiform script. The biologist's model suggests, however, that this was not so.

Wild onagers were certainly hunted for food and for sport in the ancient world (as they were until modern times), and their bones are frequently found amongst other food debris on archaeological sites of prehistoric and protohistoric date. Zeuner (1963) following the work of earlier writers, especially that of Hilzheimer on the northern Mesopotamian site of Tell Asmar (1941), believed that the osteological remains of onager were from domesticated animals. There is, however, no more evidence for domestication than there is from the remains of gazelle from the same sites. The bones do not differ from those of wild animals and there is no evidence from wear of the teeth to indicate that the onagers were ridden or driven.

Onagers were also frequently depicted in Mesopotamian art but we now believe that these were wild animals. A fine example is shown in Figure 9.13. This is a rein ring, again from Ur. It is of electrum (an alloy of silver and gold) and was also made in about 2500 BC. Many wild animals were figured on rein rings, other examples being of lions, leopards, ibex, and stags, so it is not improbable for the wild onager to be included. This particular rein ring was found in the grave of Queen Puabi (Shub-Ad) and it was part of the harness trappings that were associated with

Figure 9.13 Onager (*Equus hemionus*) on the rein ring from the grave of Queen Puabi, Iraq. Now in the British Museum (photo BM).

Figure 9.14 Slaying onagers, opposite, with arrows (*Equus hemionus*) and capturing them alive with ropes. Scenes from the Palace of Ahurbanipal, Nineveh, *c.* 645 BC (photo BM).

a wooden sled and the skeletons of two oxen (now on exhibition at the British Museum, London). It is therefore proven that the onager on this rein ring does not signify that the sled was drawn by domesticated onagers.

It should not be concluded from the example of this one burial that onagers were never used in harness but, by taking the positive evidence that is available at present, the following deductions can be made:

The donkey and the horse were both introduced as domestic animals into western Asia during the early part of the third millennium BC. Mules and other equid hybrids became common about 2500 BC. Donkeys, horses, and their hybrids were all used for riding, as pack animals, and for drawing carts, sleds, and chariots (as were cattle). Onagers were not domesticated but Zarins (1976) has suggested that they were interbred with donkeys and horses.

The Syrian onager, *E. hemionus hemippus*, inhabited alluvial plains throughout the Near East whilst the Persian onager, *E. hemionus onager*, a larger animal, lived at a higher altitude and ranged further to the east. Both subspecies played a role in the culture of the ancient civilizations and their meat was a relatively important source of food. The Assyrian hunting scenes from Ashurbanipal's palace (Fig. 9.14) show that as late as 645 BC the Syrian onager was valued as sport for kings. The animals are shown being shot at with arrows and a male onager is being caught with a rope, for what purpose is unknown.

Section II
Exploited captives

Exploited captives

The species of mammal that have been enfolded into human societies can be broadly differentiated into two groups and, although, because of overlaps and exceptions, these groups cannot always be separated very clearly, it is useful to define their underlying differences. The first group, here termed *Man-made animals* and described in Section I, comprises the mammals that have been moulded by man for his personal satisfaction and gain. The livelihood and breeding of the animals is entirely under human control and some, like certain breeds of dog, have been altered out of all recognition from the wild progenitor.

This group of mammals is sometimes called 'domestic', whilst the second group, described in Section II of this book under the heading *Exploited captives*, is distinguished as 'domesticated'. Van Gelder (1969, 1979) has defined these terms in this way: *Domestic animals* are *populations* that through direct selection by man have certain inherent morphological, physiological, or behavioural characteristics by which they differ from their ancestral stocks. *Domesticated animals*, on the other hand, are *individuals* that have been made more tractable or tame but whose breeding does not involve intentional selection.

Van Gelder is careful to emphasize in these distinctions that 'domestic' applies only to populations which are in practice reproductively isolated from the wild parent species, whilst 'domesticated' applies only to individuals. In this sense any mammal can be domesticated or tamed, but the term is mostly used for such animals as the Indian elephant that normally breeds in the wild but is then captured, tamed, and trained as a beast of burden (Fig. II.1).

The following Section on *Exploited captives* includes all those domesticated mammals whose breeding remains more under the influence of natural rather than artificial selection although the species may have been associated with man for thousands of years. The reason that there has been little human interference with the breeding of these exploited captives is because it is their perfect adaptation to a harsh environment that is of the greatest benefit to man. For example it would not be possible by artificial selection to improve on the dromedary's ability to

Illustration on pp. 102–103. Reindeer in a Lapp camp, see p. 130

survive in the Sahara and any change would be likely to be harmful both to the camels and to the desert nomads who depend on them for survival. This is not to say that selective breeding is never carried out, there are for instance special breeds of racing dromedaries and pack dromedaries, but in the main it is the natural strength and endurance of the species that is exploited. This applies to all those domesticated mammals that enable human populations to flourish in the harsh or uncongenial parts of the world, the reindeer in the Arctic, the yak in the Himalayas, and the water buffalo in the swampy rice fields of Asia.

Section II begins with a chapter on the cat, a mammal that does not really fall into either group but is intermediate between domestic and domesticated. Perhaps it should be called 'an exploiting captive'. The cat is a solitary carnivore that enjoys the company of man but is just as happy to return to Kipling's 'wet wild woods waving his wild tail and walking by his wild lone'.

Figure II.1 Two Indian elephants beneath a tree. India, Deccan region, c. AD 1660 (India Office Library).

10 Cats

ORDER CARNIVORA, FAMILY FELIDAE

He will kill mice, and he will be kind to Babies when he is in the house, just as long as they do not pull his tail too hard. But when he has done that, and between times, and when the moon gets up and night comes, he is the Cat that walks by himself, and all places are alike to him. Then he goes out to the Wet Wild Woods or up the Wet Wild Trees or on the Wet Wild Roofs, waving his wild tail and walking by his wild lone.

Kipling, in the *Just So Stories*, showed himself to be a master of assonance; he was also good at observing animal behaviour, but he was quite wrong about the cat when he said, 'all places are alike to him'.

The cat is strongly territorial in its behaviour and it is a solitary, and at least partly nocturnal, hunter. By these characteristics it is set apart from all other fully domestic animals, but on the other hand it is its territorial instincts that keep this carnivore close to man. It may be stated that by their offerings of food, affection, and comfort humans persuade cats to share the same core area of their home range (see p. 55). Nearly all domestic cats can, however, survive and even flourish on their own if they shift this core area or home base to a farm, a derelict building, or even to land that is uninhabited by man.

Cats can live as closely cherished domestic pets, or as tolerated half wild animals that inhabit cellars and outbuildings, or they can flourish as truly feral carnivores that live entirely by hunting. Probably every city and almost every farm in the world has its population of stray cats, and usually they are treated with respect by humans who cannot resist the compulsion to feed and protect these shadowy but endearing creatures. As was said in the introduction this follows from man's instincts for food-sharing and nurturing. One large public building in London, notorious for its population of stray cats which it has been unable to eradicate for the last fifty years because the employees will insist on feeding the animals, recently put up the following notice:

It is strictly forbidden to feed cats in this area. The authorised feeding place is at the north east corner of the site by the builders' skip.

And sure enough at the appointed place there is a special platform on which the food may be prepared and a very large sawdust-filled tray to receive the cats' excreta.

Considering the thousands of years over which man has been associated with the cat it is surprising how little difference there is between the wild and the domesticated forms. This is because, although the cat is a fully domesticated animal, its breeding is seldom controlled by man and it usually reproduces as a wild animal subject more to selective pressures imposed by its environment than to artificial selection by man for favoured characters. There is therefore no problem in recognizing the progenitor of the domestic cat in *Felis silvestris*.

Despite man's constant efforts to exterminate this wild cat it is still fairly common and has a ubiquitous distribution over Europe, Africa, and Asia. In both its wild and its domesticated form *Felis silvestris* must be counted as a most highly successful carnivore, and like most widespread mammals the wild species has evolved into a great many geographical races that are adapted to local conditions of climate and environment. The wild cat conforms to Allen's law in being a solidly-built carnivore with thick fur, a rather short face, short legs, ears, and tail in the northern, cold latitudes, whilst in southern Europe, and in the tropics it is a fine-limbed cat with a long face, long pointed ears, a long tail and a short sleek coat. The two extremes of this cline are so distinct that until recent years they were placed in separate species with the European wild cat being named *Felis silvestris* (Fig. 2.2) and the African and Arabian form being named *Felis libyca*. It is this southern form that has always been favoured as the main line progenitor of the domestic cat. Nowadays it is more usual to treat these two forms of wild cat as subspecies so that we have *Felis silvestris silvestris* and *Felis silvestris libyca* with the innumerable intermediate races being assigned to other subspecies, such as *Felis silvestris jordansi* for the wild cat from Majorca.

Due to the pertinacity of the taxonomist Pocock, who repeatedly reviewed the classification of the Felidae during the first half of this century, the nomenclature of the family appears to be particularly complicated and confusing, and there are more than 230 names that have been applied at different times to the species and subspecies of felids. Today the proliferation of names for geographical races in this way is frowned upon by taxonomists and it is usual to concede only about 28 species to the genus *Felis*. In one of his earlier publications Pocock (1907) examined, with commendable restraint, the probable origins and affinities of the English domestic cat, basing his views on features of skull shape and pelage. This work still stands as a factual and accurate account, and although much work has been done in recent years on the genetics of coat colour and pattern, no new results have emerged to gainsay his conclusions that all present-day breeds of domestic cat are descended from the European wild cat, *F.s. silvestris*, and the African wild cat, *F.s. libyca*. Pocock made the further supposition that it was the latter subspecies that was domesticated first and that interbreeding then took place with the northern wild cats. This theory is lent support by his assertion that when *F.s. silvestris* is interbred with *F.s. libyca* the offspring resemble the domestic striped tabby in their pelage characters more than they resemble either parent. Furthermore in the Middle East there is an intermingling between the northern and the southern forms of wild cat and in this area it is particularly difficult to distinguish the feral cat population from the wild one. It has long been maintained, although not by Pocock, that the European wild cat has light-coloured fur on the back of the lower part of the hind leg (below the hock) whilst the African wild cat has black hair on this part, and that most domestic cats conform to the pattern of the African cat. However, as in all the pelage characters of wild and domestic cats this is seen to be an inconsistent feature when a large series of skins is examined.

Figure 10.1 Striped tabby cat, '*Felis torquata*'

The great majority of domestic cats, all over the world, are tabbies; that is, like the wild cat, they have a coat that is blotchy grey with black, tawny-ochreous, and lighter coloured spots and stripes. This type of pelage, in which the individual guard hairs are banded in varying shades of grey, is often called agouti by taxonomists after the South American rodent which also has banded guard hairs. The pelage of domestic tabbies falls into two distinct patterns, neither of which is, however, exactly like that of the wild cat. These two types were described by Pocock as follows:

1. STRIPED TABBY = *Felis torquata* (see Fig. 10.1)
Sides of the body, from the shoulder to the root of the tail, marked with narrow wavy vertical stripes which show a tendency, especially on the thighs, to break up into spots; no broad latero-dorsal stripe.

2. BLOTCHED TABBY = *Felis catus* (see Fig. 10.2)
Sides of the body marked with three usually obliquely longitudinal stripes forming the so-called 'spiral', 'horseshoe', or 'circular' pattern of fanciers; a broad latero-dorsal stripe on each side of the narrow median spinal stripe.

In the old literature the name *Felis catus* is applied to the European wild cat but, as pointed out by Pocock, this is the name that Linnaeus gave to the blotched tabby domestic cat and it is therefore misleading for the wild species (see Appendix I). The situation is made more confused by the fact that the striped tabby usually resembles the pelage of the wild cat, particularly *F.s. libyca* more than does the blotched tabby.

In general it may be stated that the striped tabby domestic cat has a coat that is a combination of the characters most commonly seen in both the European and the African wild cats, but the blotched tabby is distinctly different and according to

Figure 10.2 Blotched tabby cat, '*Felis catus*'

Figure 10.3 Black cat

Todd (1978) this pelage must have arisen as a highly successful homozygous (true breeding) mutation from the striped type. Todd, in a study of the population genetics of domestic cats, examined the three most common variations on the wild pattern that are seen in the pelage and has suggested that they arose as mutations that are related to behavioural traits and are subject to differential selective pressures in urban and rural environments. The three variations from the striped tabby (that Todd equates with the wild or agouti type) are the blotched tabby, the non-agouti or black, and the sex-linked orange, Figure 10.3. For his study Todd sampled populations of cats in cities throughout Europe, North Africa, and western Asia and he then drew clinal maps to demonstrate the distributions of the four varieties. From these clinal maps, of which Figure 2.6 is one, he has postulated the focus or probable place of origin of the mutants. Whether or not these predictions are true the study exemplifies how natural selection working across an urban–rural axis can affect populations of animals that are fully domesticated but whose breeding is outside man's control. It also helps to explain how the cat has become such a remarkably successful animal in both the countryside and city. It seems that it not only has nine lives but the cat also has a different coloured coat and different character depending on where it lives.

There is little or no information from the literature or early pictorial representations to indicate how ancient the four main groups of cats are; these being the two varieties of tabby, the single-coloured black or white, and the sex-linked orange (marmalade and tortoiseshell cats). In addition there are of course other breeds of cat that are more closely controlled by man, such as the Manx, the Persian, Siamese, and Abyssinian, to name but a few. These breeds are unlikely to be of ancient origin, although their history is little known, and they fall outside the range of this book.

One aspect of the history of the domestic cat that should be mentioned is how far species, other than *Felis silvestris*, have been interbred with domestic stock or tamed. For example, it has been suggested that the Persian long-haired cats are descended from Pallas's cat, *Felis manul*, a wild cat that inhabits central Asia and which is unmarked with spots or stripes and has very long soft fur. There is, however, no osteological or other evidence for this and it is more likely that the long-haired domestic cats are the result of artificial selection for this character by man.

Another species of wild cat that has sometimes been accredited with partial ancestry of the domestic cat, especially those in India, is the jungle cat, *Felis chaus*. This cat has a uniformly coloured body, being a tawny grey with darker specks or ticks, a dark banded, rather short tail and long tufted ears. The jungle cat is much larger than *Felis silvestris*. It ranges from Egypt across south west Asia, and from Afghanistan into India and east to China and south east Asia.

A small collection of skins of domestic cats from India that is in the British Museum (Natural History) was described, with some reservations, by Pocock (1951) as being of the 'chaus-type'. The pelage in these specimens is grey agouti and the only tabby markings are on the legs and at the end of the tail. Blyth, who collected a few of these cats in the 1860s believed that they were feral animals that resulted from domestic cats that had bred with wild *Felis chaus*. There is, however, no real evidence for the true domestication of the jungle cat, although the corpses of this species were mummified by the Ancient Egyptians along with those of many other wild and domestic animals. It is therefore likely that the Egyptians tamed the jungle cat and may have interbred it with domesticated *F.s. libyca* (the hybrid offspring are said by Gray, 1972, to be fertile), but it is unlikely that occasional crosses between these two species have had much impact on the stock of domestic cats, outside Egypt and perhaps India.

In Figure 10.4 can be seen specimens from a collection of 190 cat skulls that were excavated by Flinders Petrie from tombs at Gizeh in Egypt, at the beginning of this century, and presented by him to the British Museum (Natural History). The skulls are from mummified animals (the skeletons, mandibles and wrappings were presumably, unfortunately, thrown away), and they probably date from the late first millennium BC. Three of the skulls have been identified as belonging to *Felis chaus* by Morrison-Scott (1952) and the rest have been assumed to belong to domestic cats although the skulls were not distinguishable from those of wild *F.s. libyca*. This is not surprising, however, as it is often difficult to separate the skulls of present-day domestic cats from those of the wild parent species.

The cat was one of the most sacred of all animals to the Ancient Egyptians and they were mummified in enormous numbers (Fig. 10.5). It was absolutely forbidden to kill a cat and Herodotus tells how if one died of natural causes in a house, all the members of the household shaved their eyebrows. When the animals died they were taken to the city of Bubastis where they were embalmed, after which they were buried in sacred repositories. Huge collections of these mummies were removed by excavators at the beginning of this century and so great were their numbers that they were spread on the ground as fertilizer. One cat skull in the collections of the British Museum (Natural History) (separate from the 190 described above) is all that remains from a consignment of 19 tons that was shipped to England to be ground up for this purpose; a practice that was equally vandalous, if rather more ingenious, than the present one of depleting the seas of fish for the manufacture of fertilizers from fish meal.

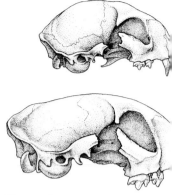

Figure 10.4 *Felis silvestris libyca*, above, and *Felis chaus*. Skulls from cats that were mummified in Ancient Egypt.

Figure 10.5 Mummified cat from Ancient Egypt, unprovenanced (photo BM(NH)).

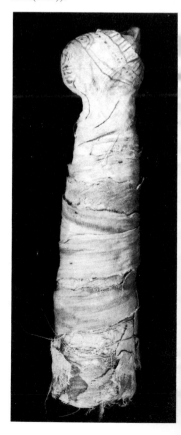

It is not known when cats were first domesticated. Zeuner (1963) believed that it was comparatively late, although in an earlier publication (1958) he clearly liked the idea of suggesting that the remains of cats from the Pre-pottery Neolithic of Jericho (*c.* 7000 BC) were from animals that had some sort of association with man. There is little pictorial evidence from Ancient Egypt to indicate that the cat was fully domesticated until the New Kingdom (Eighteenth Dynasty, *c.* 1600 BC) (Fig. 10.6, p. 79). However, it may well be that cats were tamed and lived in association with humans very much earlier than the archaeological and historical records imply. Their remains are fairly frequently retrieved from prehistoric sites along with those of other wild carnivores such as foxes, otters, and badgers, but it is seldom possible to assess whether these were animals that were pets or were killed for food or, more probably, for their pelts.

The relationship between cats and people is perhaps more symbiotic and mutually beneficial than that commonly found between any other animal and man. The domesticated cat extracts from its human partner a home, warmth, affection, and play, whilst the human can assuage his natural inclination to nurture a warm, soft, furry animal and at the same time his household will be kept cleared of unwanted scavenging animals such as rats and mice. There can be little doubt that this association is a very ancient one, but there is another side to the relationship, this being the part played by cats in witchcraft, sorcery, and various cult practices. Presumably because it is a nocturnal and solitary hunter that can make the most blood-curdling cries when it is fighting, the cat has, more than any other animal, been associated with superstition and sympathetic magic. In Britain one such superstition for which we have the material evidence

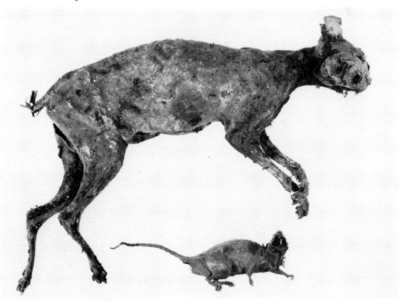

Figure 10.7   Dried cat and rat that had been built into a wall together. Found when a house was demolished in Bloomsbury, London (photo BM(NH)).

was the belief that if the corpse of a cat was built into the walls of a building, rats and mice would be kept away. This theory must have been held particularly by the builders of the Blooms-bury region of London as late as the eighteenth century AD, for several dried cats have been found in the walls of houses during re-building within recent years. Sometimes presumably this happens even today by mistake when cats crawl into a half-built wall or roof and become trapped by the builders unknowingly, but very often there is evidence that the cat had been killed, and placed in a life-like position, sometimes with a rat beside it or even in its jaws (Fig. 10.7).

11 Elephants

Figure 11.1 Reconstruction of a mammoth, *Mammuthus primigenius*

Figure 11.2 Reconstruction of a straight-tusked elephant, *Palaeoloxodon antiquus*

* Terminology used in British Isles; see Appendix II for correlations with other regions.

ORDER PROBOSCIDEA, FAMILY ELEPHANTIDAE

As with many groups of large mammals the elephants of today are the relics of a once much more diverse family that during the Pleistocene inhabited every part of the earth with the exception of Australasia and Antarctica. The association of hominids with elephants stretches back into the Middle Pleistocene, perhaps 400 000 years ago. By this period the ancestral southern elephant, *Archidiskodon meridionalis*, had moved north into Europe and had evolved into two very distinct forms, the woolly mammoth, *Mammuthus primigenius*, that inhabited the steppe-tundra of the frozen north and the straight-tusked elephant, *Palaeoloxodon antiquus*, that inhabited Spain, and the warmer regions of the north during interglacial phases. Both elephants were extensively hunted by early man and it is probable that from their beginnings hominids, competing only with the large cats, were these elephants' main foe.

In the northern hemisphere the straight-tusked elephant and the mammoth were both exterminated, the former at the end of the Hoxnian Interglacial* (see chart, Appendix II, p. 199), whilst the mammoth survived until the end of the last glaciation (about 11 000 years ago) (Fig. 11.1). In the southern hemisphere elephants belonging to the Recent genera, *Loxodonta* and *Elephas* also evolved during the Pleistocene and they too were preyed upon by hominid hunters, as well as by the lion and tiger. Why the straight-tusked elephant and the mammoth became extinct whilst the southern genera survived and flourished to evolve into the living species was probably mainly a result of climatic and environmental conditions that changed much more drastically and rapidly in the north, during the Ice Age, than in the south. Although human hunters in Europe and Asia may have been more prolific and more efficient at killing than those of the southern continents during the late Pleistocene it is unlikely that alone they could have had a lethal impact on the populations of elephants. The straight-tusked elephant (Fig. 11.2) was driven south, along with the hippopotamus and the rhinoceros, with the advancing ice sheets of the Wolstonian* (see chart, Appendix II), whilst the mammoth was driven north with the final retreat of the

113

ice at the end of the last glaciation (Devensian*). There is good evidence to indicate that its end was hastened by man but this migrating, arctic giant might not in any case have survived the 'climatic optimum' of the early Holocene. There is, however, much controversy on this question which has been reviewed by Martin & Wright (1967).

As far as is known at present from the subfossil and archaeological evidence the only elephants to survive the end of the Ice Age were the present African and Indian forms. These are classified as *Loxodonta africana* and *Elephas maximus*, these names being a good example of the anomalies that can happen in nomenclature for it is the African elephant that is usually the larger of the two. The rules of nomenclature, however, insist that the earliest name used to describe an animal or plant must prevail and Linnaeus who first named the Indian elephant may have been following the classical writers who firmly believed that the Indian elephant was the larger.

Although there is no osteological evidence for the survival into the Holocene of any other species of elephant it is possible that on some islands dwarf forms may have lived on until the arrival of man. The remains of what would seem to us tiny elephants, only a metre high at the shoulder, have been retrieved from Pleistocene deposits on Malta and Cyprus, and others have been recorded from Java, Sumatra, and the Santa Barbara Islands off the coast of California.

There is a certain amount of literary and pictorial evidence to indicate that wild elephants existed in the Near East, particularly in Syria, in early historic times. It is usually believed that these elephants belonged to the Indian species, *Elephas maximus*, and certainly these were imported into western Asia in classical times, but the lack of a fossil record leaves the evidence insubstantial and hazy. If there were truly wild, indigenous elephants in Syria, then it is possible on zoogeographical grounds that they were related to the Indian form, but it is equally likely that they were of African origin, or they may even have evolved separately and have been distinct from both living groups (Fig. 11.3).

The elephant of the African savanna is not only larger in overall body size than the Indian it also has larger ears and both sexes carry tusks which may be exceptionally long and heavy in the male. The profile of the back of the African elephant is straight or sway-backed whilst in the Indian it is convex or domed and in Asiatic elephants only the male carries tusks as a rule. If a female Indian elephant has tusks they are small and inconspicuous. The enamel plates of the molar teeth are greater in number and therefore closer together in the Indian elephant (Fig. 11.4).

The African and Indian elephants are said to differ in the way that they use their trunks and feet. In the African, the tip of

* See footnote, p. 113.

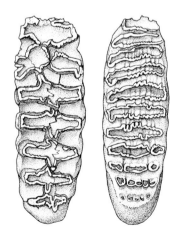

Figure 11.3 African elephant, *Loxodonta africana*, above, and Indian elephant, *Elephas maximus*

Figure 11.4 Molar teeth of an African, left, and an Indian elephant

the trunk has two finger-like processes, whilst there is only one in the Indian and it is said that these elephants are more agile at using their feet in conjunction with the trunk for breaking branches or moving boulders. African elephants, according to Sikes (1971), rarely use their forefeet for any operation other than digging or scraping soil. All elephants are highly social mammals that will remain together as an integrated family group over several generations.

Both genera of living elephants are usually held to contain only a single species but several geographical races have been described for each. There is one race of African elephants, *Loxodonta africana cyclotis*, that is of particular interest in the context of human exploitation. This is the so-called forest elephant and it is rather smaller than the savanna or bush elephant which is today more common (Fig. 11.5). It has a shoulder height of less than 2·4 metres, but it is not a pygmy race as is often claimed in the older literature. The forest elephant has smaller, more rounded ears than the commoner savanna form and the tusks are smaller and straighter; the ivory is said by craftsmen to be 'harder'. Today this race is found only in isolated parts of west Africa and west-central Africa across to western Uganda but it could well be that it was this elephant that once ranged all over North Africa and was domesticated and finally exterminated north of the Sahara by the Romans. If this were so it would explain why the classical writers believed the Indian elephant to be the larger, and it is even possible that it was this race that was indigenous in western Asia.

The names 'forest' and 'savanna' elephants are misnomers because all elephants, both African and Indian, are browsers whose preferred habitats are luxurious, wooded jungles, and forests with plenty of water. Like most other large mammals, however, they are versatile in their tolerance of temperature, humidity, and foodstuffs, and they can survive in any environment that can provide them with enough vegetation for sustenance.

The interaction of humans with elephants may be divided into three categories: firstly, predation for food which was presumably the earliest form of exploitation. Secondly the killing of elephants for their ivory alone, a trade that has been of the greatest importance in human economies since the rise of the earliest civilizations. Finally there is the taming of live elephants for use in warfare, in circuses and zoos, and as beasts of burden, and this has a history of at least 4000 years.

Except when it happens in zoos, tamed elephants have never been bred in captivity over many generations and subjected to artificial selection as has been the practice with other domesticated animals. This is because it is uneconomic and also because it is probably not possible to 'improve' on the elephant's use to man;

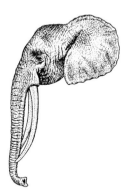

Fig. 11.5 African forest elephant, *Loxodonta africana cyclotis*

its powers of strength and endurance are already quite sufficient. In addition the mating behaviour of elephants is difficult to control, there is a very long gestation period (22 months), and like humans they are not adult until into their 'teens. Both African and Indian elephants are highly intelligent and responsive to human command. Indian elephants have probably been tamed since the time of the Indus Valley civilizations of Mohenjo-Daro and Harappa (*c.* 2000 BC) for they are on steatite seals from these sites, sometimes with a cloth on the back of the animal indicating its use to man. African elephants are not commonly tamed at the present day and it is often said that they are much more difficult to train than the Indian. The Romans used them extensively, however, both for warfare and in their circuses, although as has already been said, probably it was the smaller race of *L.a. cyclotis* that was known in the classical world.

ELEPHANT HUNTERS

Ice-age occupation sites in the Ukraine (see the map, Fig. 11.6) have provided abundant evidence that Upper Palaeolithic people not only hunted mammoths but used their bones and tusks in the construction of their dwellings. These have been described (for English-speaking readers) with great clarity by Klein (1973, 1974) who states that 'mammoth bones are virtually the hallmark of these ice-age sites'. Klein maintains, however, that there is little evidence for the inhabitants, who lived between 75 000 and 10 000 years ago, actually killing mammoths. He believes that the hunters collected limb bones and tusks from the bare and treeless landscape to use as building material because of the scarcity of wood. Klein's reasons for assuming that the mammoths were not killed by man rests on the dating of the bones which has shown that there may have been thousands of years between the death of the animals and the habitation of the sites. It is very likely true that the inhabitants of these particular sites may not have killed the mammoths whose bones they collected but this does not preclude their slaughter by man initially, even if it was millennia before-hand. The bones would then have been gnawed by scavengers, such as hyaenas, and at any later date when they had been cleaned and dried by the processes of nature, they would provide admirably suitable material, either for building or for ritual purposes and carvings.

The drawing of a reconstruction of an ice-age hunters' hut from the Ukraine may be compared with that of an actual elephant hunters' hut in the Ituri Forest of Kinshasa (Fig. 11.7). In this region of Africa elephants have always been important to man as suppliers of meat and raw materials, and unlike the profligate ravages of the European 'big game hunters' when an elephant was killed by the pigmies in the Ituri Forest nothing was wasted. The dried meat from one elephant could supply a small village with valuable protein for several weeks, and there

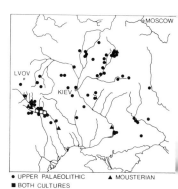

Figure 11.6 Archaeological sites in the Ukraine that have provided evidence for the hunting of mammoths (from Klein, 1974).

Figure 11.7 Reconstruction of a mammoth hunters' camp in the Ukraine (from Klein, 1974) opposite, compared with that of a modern elephant hunters' camp in the Ituri forest, Zaire, below (from Sikes, 1971).

were many complicated rituals and customs that accompanied the no mean task of killing an elephant before the advent of firearms. This was accomplished with the aid of traps, pitfalls, poisoned arrows, and snares.

THE IVORY TRADE

There appears to be little information on who killed the elephants that provided the vast amount of ivory that was used in the Ancient World, nor how the export of ivory was organized. To Homer, the Greek word *Elephas* meant ivory not the animal, and even Herodotus although he mentioned the elephant in his writings never saw one, and yet ivory had been one of the most important and valuable of all raw materials for many millennia before the time of Homer. As has been said with reference to sites in the Ukraine, Palaeolithic man in the Pleistocene used ivory for artefacts, for carvings, and as a building material. Coming down into the Recent era, excavations at Knossos, the Minoan city on Crete (*c.* 1400 BC), have provided evidence for an ivory worker's shop and four elephant tusks were found at the Minoan palace of Zacro, whilst Mycenae on the mainland of Greece has produced a carved elephant tusk. After this period when ivory working was common it seems to have become rather less popular outside Egypt, until the ninth and eighth centuries BC when vast quantities of ivory were used in the decorations of the palaces of Mesopotamia, as well as in the manufacture of jewellery and ornaments. Amongst the most famous of ancient ivories are the Assyrian carvings from the palace of Nimrud that are now in the British Museum, London (Fig. 11.8). Some of these were obtained by Sir Henry Layard as early as 1845 and the rest were excavated by Mallowan, more than a hundred years later. It is not known how much of this ivory came from the local Syrian elephant (assuming that wild elephants really did inhabit western Asia) and how much came from Africa and India, and it would be impossible to assess this from the structure of the ivories which has deteriorated from being buried for two thousand years. Although it is often stated that it is possible to distinguish the ivory of African and Indian elephants, recent tests at the British Museum (Natural History), including use of the scanning electron microscope, have failed to substantiate these claims, and Mesopotamian ivory has no special characteristics that would differentiate it (Fig. 11.9).

With the Roman Empire the demand for ivory increased to the

Figure 11.8   Carvings in ivory from the palace of Nimrud (Iraq): Lioness attacking a negress, eighth century BC; Sphinx, ninth to eighth centuries BC; Ivory or Phoenician, see p. 120 (photo BM).

point of extreme extravagance, the Romans being undeniably responsible for the extinction of the elephant in North Africa. Caligula gave his horse an ivory stable, whilst Seneca had five hundred tripod-tables with ivory legs (Scullard, 1974). The ivory trade at this period must have rivalled that of the British Empire at its peak during the nineteenth century when every European traveller wished to shoot an elephant and the demand for ivory for buttons, billiard balls, knife handles, and piano keys was unlimited.

Craftsmen in ivory, before the age of plastics, could assess the quality of ivory by its appearance, colour, and feel. The best ivory for billiard balls was obtained from West African elephants. East African elephants produced 'soft' ivory and Indian elephants had tusks that, although they are fine-grained, were not so highly esteemed as the African. Another source of ivory that was exploited to a surprisingly large extent was that from frozen mammoth tusks found in Siberia and Alaska. Mammoth ivory has been known about and used since at least the Medieval period, about 15 per cent of that retrieved being of high quality, although it is often of a golden yellow colour.

Eighteenth century travellers in the New Siberian islands of the Arctic Ocean described how when the ice-covered sand cliffs were thawed by the summer sun the surface would slip down, bearing with it great quantities of mammoth bones and tusks. In 1821, 90 000 kg of ivory were taken from these islands and the supply was thought to be inexhaustible.

Figure 11.9   Detail from the 'Black Obelisk', an Assyrian stone monument inscribed with the annals of Shalmaneser III (858–828 BC) and representing the tribute sent to him by subject princes. From Nimrud, now in the British Museum. Maybe this is a Syrian elephant (photo BM).

Figure 11.10 Ancient coins depicting elephants:
a) silver tetradrachm of Seleucus I, Alexander's successor in Syria, *c.* 306 BC.
b) Tetradrachm of Antiochus I.
c) Alexander the Great attacking Porus on an elephant, decadrachm minted in Babylon. (photos P. Clayton).

DOMESTICATION OF THE ELEPHANT

Live elephants have been caught and tamed by man for more than 3000 years. They are captured by means of elaborate pitfalls, lassoos, or by being driven into stockades with the help of tamed animals used as decoys. Or, alternatively, there is the wasteful method of killing the adults in a herd in order to capture the infants. Once under the dominance of humans, elephants, both African and Indian, can be relatively easily tamed and appear to enjoy learning new activities, even when they are adult, but their uses are limited.

Within recent times elephants have been used in the timber industry in the Far East but in the Ancient World a team of human slaves was probably more manageable and more efficient than an elephant for all its great strength. Elephants, were, however, a curiosity to be exploited in the Roman circuses and they were a symbol of power in war. In 331 BC Alexander the Great fought with the Persian king Darius at the Battle of Gaugemela, east of the river Tigris (Fig. 11.10). Alexander is said to have had an army of 7000 men, whilst Darius had perhaps 30 000 including men from Afghanistan and India who had 15 elephants. Despite such a display of strength Alexander won this decisive battle, and so it was to be in future battles, for elephants were not made for fighting human wars. Despite this seemingly obvious fact both Indian and African elephants were frequently used for front-line attacks as well as for carrying baggage. The use of tamed African elephants by the Carthaginians and by Hannibal in the Punic Wars against the Romans is very well documented and has been admirably described by Scullard (1974). The short term success of the Carthaginian 'cavalry' was probably due as much to the effect of surprise as to the power of the elephants for the one great disadvantage of an elephant in war is that if assailed by a multitude of arrows it will very sensibly turn round and go backwards, thereby inflicting worse damage on its own army than on the enemy.

Elephants were used mostly by the Romans as a public spectacle. They were made to fight with humans and lions in the Games and were trained to carry out the most skilful tricks in the circuses, as they are today. Pliny describes how elephants could throw weapons in the air and play at a war dance. They learnt to walk on 'tightropes', and would pick their way amongst seated people, with great care, to take their places at a banquet.

The elephant is probably one of the very few species of mammal whose use to man as a domesticated animal is drawing to a close. As a symbol of power and for pageantry in the Ancient World and in the Far East the elephant has no peer but there can be little future for these roles in the modern world of machinery. It may be questioned what relevance the hunting of mammoths and the ivory trade has to do with the history of domestication, but the elephant is unlike other domesticated animals in that its exploitation by man has not conformed to the usual pattern of taming, and artificial selection following breeding in captivity. Elephants have never been enfolded within human society like the horse or the pig have been. Nevertheless elephants of all races have been closely involved with humans for perhaps half a million years and it will be mankind's loss if we continue the process of extermination that was begun by Palaeolithic man on the mammoth.

Figure 11.8, see p. 117

12 *Camels and llamas*

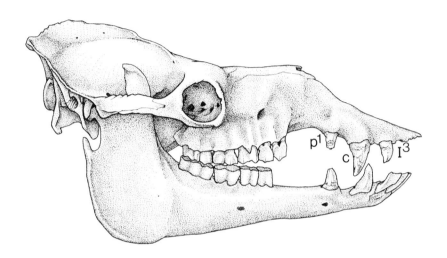

Figure 12.1   Skull of a camel. Skull length 480 mm

ORDER ARTIODACTYLA, FAMILY CAMELIDAE

Camelids are even-toed ungulates without horns. There are six species living today, these being the one and two humped domestic camels of Africa and Asia, and the guanaco, llama, alpaca, and vicuna of South America. Taxonomically the camelids fall between the pigs and the true ruminants and they are classified in the suborder Tylopoda. The Camelidae differ from true ruminants in a number of important characters. Firstly the teeth are rather different in that there may be traces of vestigial central incisors in the upper jaw, or premaxilla, and the third incisors are well developed as permanent, canine-like tusks (Fig. 12.1). In addition there are true canine teeth, or tusks, in both jaws, and posterior to these the first premolars are also developed as tusk like teeth which are separated from the grinding cheek teeth by a gap, or diastema. Secondly the musculature of the hind limbs is different from that of other ungulates in that the legs are very long and are attached to the body at the top of the thigh only; in horses and cattle the legs are attached by skin and muscle from the knee upwards. This means that when a camel lies down it can rest on its knees with its legs tucked under its body in quite a different way from that of cattle (Fig. 12.2). The feet do not have functional hooves, the toe bones being embedded in a broad cutaneous pad on which the

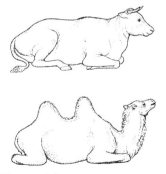

Figure 12.2

121

Figure 12.5 A herd of dromedaries grazing in scrub desert, Iraq, see p. 124 (photo author).

Figure 13.1 Reindeer in´a Lapp camp, Finnmark, Norway, see p. 130 (photo John Berge).

Figure 13.3 White reindeer, Finnmark, Norway, see p. 130 (photo Johan Berge).

Figure 12.3   Camel's foot-pad

Figure 12.4a   The four living New World camelid species

Guanaco, *Lama guanicoe*

animal walks (Fig. 12.3). This pad is more highly developed in the Old World camels than in the South American species. Although the camelids do ruminate, that is they regurgitate their food and chew a cud, the stomach has a different construction from that of the true ruminants. The camelids of both the New and the Old Worlds are adapted in their anatomy and in their physiology for survival in particularly harsh desert environments and high altitudes.

There are generally recognized to be three genera and six species of living camelids (Fig. 12.4). These are listed as follows:

THE NEW WORLD CAMELIDS

South American camelids which today, as well as within historical times, have been confined to the western and southernmost parts of the South American continent.

Lama guanicoe, the guanaco, is a wild species and is the tallest of the New World camelids. It is most common in semi-desert and high altitude plains but ranges from the highlands of the Andes in Ecuador and Peru to the plains of Patagonia.

Lama glama, the llama, is only known as a domestic animal. Llamas are common in the mountains of Peru where they are the most important all-purpose animal.

Lama pacos, the alpaca, is also known only as a domestic animal. It is extremely shaggy and is the chief provider of wool in Peru. The alpaca is smaller than the llama and is not used as a pack animal. The fleece may be black, brown, or white and is sheared every two years so that each animal is only shorn three or four times in its life and yields about 3 kg of wool each time. Alpacas live most successfully at altitudes higher than 3600 m.

Vicugna vicugna, the vicuna, is the smallest of the camelids and is a wild species inhabiting the Andes between 4000 and 5000 m. Vicunas live close to the snow line and they have a coat of wool that is probably the finest and lightest in the world. This wool is obtained by driving the vicunas into corrals, shearing them and then releasing them again. The vicuna has never been domesticated.

Llama, *Lama glama*

Alpaca, *Lama pacos*

Vicuna, *Vicugna vicugna*

The four South American camelids all have a diploid chromosome number of $2n = 74$ (see footnote p. 53), and they will all interbreed to produce fertile offspring.

Camelus bactrianus, the Bactrian camel, is the domestic camel with two humps that is found in the cold desert regions of Central Asia. It is possible that a small population of truly wild Bactrian camels has survived on the Mongolian side of the Altai mountains.

Camelus dromedarius, the Arabian camel or dromedary, has one hump and is more fine-limbed and runs more swiftly than the Bactrian. It is the domestic camel of the hot deserts of North Africa and western Asia and is unknown as a wild animal.

The Bactrian camel and the dromedary have a diploid chromosome number of $2n = 70$. The gestation period is 13 months in the Bactrian and 11 months in the dromedary. The species will interbreed but the male offspring are usually sterile whilst the females are fertile.

Poor camels, they have had a hard time from the hands of writers. Take for example Palgrave*:

He takes no heed of the rider . . . walks straight on when once set agoing, merely because he is too stupid to turn aside, and then, should some tempting thorn or green branch allure him out of the path, continues to walk on in the new direction simply because he is too dull to turn back into the right road . . . Neither attachment nor even habit impress him; never tame, though not wide-awake enough to be exactly wild.

Like most nineteenth century British travellers this writer lived in a world and an intellectual atmosphere that was imbued with the superiority of man. It was not possible for such people to concede that they might share any aspect of social structure or behaviour with the animal world and yet they would never hesitate to attribute the worst aspects of human nature to the beasts over whom they ruled.

Nowadays we can look at the seemingly obstinate and aggressive behaviour of the camel and interpret it in relation to its true nature, that of adaptation to an environment which necessitates the conservation of every bodily resource. In a desert or at the top of a mountain, survival depends on the ability to withstand great alternations of temperature, a restricted diet, and less water than is available in any other environment. What appears to be inflexible behaviour in the camelids is an important part of their strategy for survival, and it is this strategy that is exploited by humans who share the same environment (Fig. 12.5, p. 122).

Gauthier-Pilters (1974), writing about the camels of the western Sahara, claims it is hard to visualize an association between man and animals more close than that between the nomad and his camels. These people live an existence that is 'parasitic' on their herds of Arabian (one-humped) camels, whilst the camels depend

* Quoted from Flower & Lydekker, 1891.

Bactrian camel, *Camelus bactrianus*

Dromedary, *Camelus dromedarius*

Figure 12.4b   The two living Old World camelid species

on the nomads for their water which has to be drawn from the desert wells. A foreign traveller may think of a camel as an unpleasant animal that smells and which will bite, and spit an obnoxious fluid at passers by. A biologist studying the behaviour of the animal, or the nomads who live in harmony with the desert, know that the camel will maintain a home range, will travel along long-used, familiar trails, and will return to wells with which it has been familiar in the past. In some parts of the Sahara camels are often managed in a semi-wild state and may remain free for the four to five summer months of each year. They will return to the same wells for water at this period and the nomads may also keep them nearby with salt licks. As with other livestock animals camels that are living free are much healthier than those that are constrained because they are able to feed at night.

Male camels will not tolerate the presence of other males in the mating season if there are females in the vicinity and they will fight to the death if allowed to do so. For this reason it is usual to guard the herds well during the rutting season in the winter and surplus males are castrated. During the mating season there are three types of herds – those with one male and up to thirty females plus their one year and two year old offspring; bachelor herds consisting of males; and herds of females with small foals. The breeding male keeps to the rear of the herd where he watches his females and keeps them together in one group. Female camels breed every second year (Gauthier-Pilters, 1974).

This social structure is very similar to that seen in other domestic livestock animals such as the horse, where there is characteristically a leading male who guards his herd within a system of dominance hierarchies, and who maintains a home range but is not markedly territorial. Quite different behaviour has been observed in the vicuna. This species has never been domesticated; its behaviour and ecology have been studied in the Pampa Galeras National Vicuna Reserve in southern Peru which was established in 1966 following the realization that the vicuna would soon be extinct if it were not protected. Vicuna live in family groups within socially isolated territories that are closely guarded by the dominant male who will resist all intruders and who will expel juvenile males as soon as the maternal protection wanes. Juvenile females are also expelled from the family group and will be driven off the territory by both parents before the birth of new offspring.

Unlike other camelids the vicuna will defecate and urinate on regularly used dung piles which are used as markers for territories and as scent marks. In a high altitude or desert environment, destitute of visual landmarks, scent marks may be a necessary alternative and in addition with territorial animals the piles of dung keep intruders away. It is notable that the wild ass, *Equus asinus*, which is another desert-living territorial ungulate, also accumulates dung piles, whilst the horse does not, or at least not to the extent that is found with the ass.

ANCESTRY OF THE DOMESTICATED CAMELIDS

Extraordinarily little is known about the progenitors of either the Old or the New World camelids. The camel family first evolved in North America, as far as can be judged from the fossil record, perhaps some 40 million years ago. During the Pleistocene some species travelled across the isthmus of Panama and evolved into the living genera of *Lama* and *Vicugna*, whilst other species moved north across the Bering Straits from Alaska into Siberia. These species must have moved south and west down into southern Asia and North Africa, but few fossils have been found to trace their progress. Pleistocene camels have been identified from the Siwalik hills in India and from rather later periods on Upper Pleistocene sites in western Asia and North Africa, but the finds are rare and scattered. It cannot be ascertained how many humps, if any, these camels had.

There are wild camels living today in the western Gobi Desert, in two areas, near the Lop Nor and in south western Mongolia. The Lop Nor population is rare whilst the camels in Mongolia are said to be increasing in numbers. These camels have two humps and are named *Camelus ferus*, the assumption thereby being made that this is a truly wild species that is the progenitor of the domestic Bactrian camel. This has not, however, been proved for these camels could be descendants of domestic animals that have returned to the wild and are therefore feral. There are no living populations of one humped camels in the wild but there is some evidence from the classical literature that they existed in western Asia in the first millennium BC.

Because the camels and llama are exploited by man for their exceptional powers of endurance in a harsh environment there has been little selective breeding away from the wild form. It is therefore very difficult if not impossible to determine from osteological remains when camelids were first domesticated. This is more likely to be ascertained from the presence of figurines on an archaeological site or from the retrieval of camel dung in a cultural context than from a study of the bones. Camel dung was identified from the site of Shar-i Sokhta in central Iran (see the map, Fig. 12.6) which has been dated to approximately 2600 BC (Compagnoni & Tosi, 1978). The authors believe that the camels at this site were domesticated and that they were more likely to be the Bactrian than the Arabian species. This is the earliest evidence for domestic camels that has so far been documented.

Recent archaeological evidence from preceramic sites in the Puna of Junín in Peru has suggested that South American camelids were exploited and probably domesticated in the New World before the camel was brought under man's control in

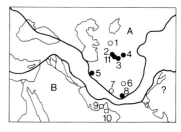

A   *Camelus bactrianus*
B   *Camelus dromedarius*

1   Tuiek-kicidzik
2   Anau
3   Altyn depe
4   Cong depe
5   Siyalk
6   Shahr-i Sokhta
7   Tepe Yahya
8   Khurāb
9   Umm an-Nar
10   Hili
11   Sor depe

●   *Camelus bactrianus*
○   *Camelus* cf. *bactrianus*
□   *Camelus dromedarius*

Figure 12.6   Possible distribution of the dromedary and Bactrian camel at the time of their domestication in western Asia, *c.* third millennium BC (from Compagnoni & Tosi, 1978).

western Asia. Pires-Ferreira *et al.* (1976) have provided a model (see below) of the way in which camelids could have been herded and domesticated in the region of the Uchcumachay Cave in the Puna of Junín, suggesting that they were first hunted and then increasingly controlled until the human population depended almost entirely on camelids for their subsistence, using them for food, wool, and draught. The authors were unable to decide whether the species represented were the llama, the alpaca, or both, but they believe that the preceramic sites of the Puna of Junín were a centre for camelid domestication beginning perhaps as early as 5500 BC.

SPANISH INVASION
Introduction of sheep
AD 1532–3
↑

Appearance of distinct
breeds of
domestic Camelidae
↑

HERDING OF DOMESTIC CAMELIDAE
↑

Increasing human control over
breeding in semi-domesticated
Camelidae
4200–2500 BC
↑

CONTROL OF SEMI-DOMESTICATED CAMELIDAE
5500–4200 BC
↑

Increasing human control over
camelid territories
↑

SPECIALIZED HUNTING OF CAMELIDAE
↑

Increasing knowledge of camelid
territorial and social behaviour
↑

Little is known of the origins of the domestic llama and alpaca. At one time it was believed that the llama was descended from the wild guanaco and the alpaca from the vicuna. After a long study on the skeletal characters and wool of the different species Herre (1952) then put forward the theory that both the llama and the alpaca were descended from the wild guanaco.

Later work on the behaviour of the South American camelids, and the osteological work on subfossil remains by Pires-Ferreira and others, suggest that the domestic camelids may be unrelated to the wild species living today. These authors postulate that there were a number of other species of wild camelid living in South America in the early Holocene and that it was these, now extinct species, that were the progenitors of the llama and alpaca. It is not unlikely that there were other wild camelids that became extinct, perhaps from overhunting, but whether they were ancestral to the present domesticated animals remains problematical and the question cannot be resolved until more work is carried out on all aspects of the South American camelids. The situation is strangely parallel to that of the Old World camels whose ancestry is also still unknown.

Although the origins of the South American camelids remain unresolved what does seem clear is that the vicuna played no part in the ancestry of the domestic species. The vicuna is placed in a separate genus, *Vicugna*, and its territorial behaviour helps to explain why this camelid was never domesticated although its wool is of exceptionally fine quality. Although a wild animal the vicuna may have been saved from extinction because the animals were valued for their wool. It is known that the ancient civilization of the Incas obtained this wool by driving the vicunas into corrals, shearing them and then releasing the animals alive. In the same way vicuna wool is obtained at the present day but the vicuna has also been extensively hunted and is now dangerously low in numbers.

Figure 12.7   Camel in warfare from the Palace of Ashurbanipal, Nineveh, *c.* 645 BC, now in the British Museum (photo BM).

The only camelid that remains as a reasonably successful wild species is the guanaco which is fairly widespread in its distribution and has greater flexibility in its feeding habits than the vicuna.

CAMELS IN PREHISTORIC AND EARLY HISTORIC EURASIA

Like the elephant, camels were domesticated for use as pack animals, and later certain societies became, and still are, totally dependent on them for food, wool, milk, draught, and dung which is used as fuel. To a limited extent they were also used in warfare, where again like the elephant their surprise effect could cause panic amongst enemy cavalry, for horses will bolt away from the sight and smell of camels if they are not used to them (Fig. 12.7).

In early January, 330 BC Alexander the Great marched into Persepolis, the ceremonial centre of the Persian Empire, with his 60 000 troops, and sacked the city. Ten thousand baggage animals and 5000 Bactrian camels had been ordered from Susa to carry away the treasure. A hundred or so years before this time Herodotus mentions camels several times in his descriptions of the fighting between the Greeks and the Persians. He describes for example how Cyrus, the founder of the Persian empire, saw the Lydians, who were on the side of the Greeks, arranging themselves for battle and alarmed at the vast size of the enemy cavalry he ordered all the camels that had come in the train of his own army to be unloaded of their baggage and to be saddled and mounted with riders. He then commanded the camels to advance on the Lydian horsemen, and as Herodotus[*] says:

> ... the Lydian war-horses, seeing and smelling the camels, turned round and galloped off; and so it came to pass that all Croesus's hopes withered away.   (I, 80)

Camels were used as baggage animals by the Romans and they were even brought into western Europe, for a cervical vertebra from a Roman villa at Soissons in northern France has recently been identified by Poplin (pers. comm.). It is very doubtful, however, if the camel would have survived long in this north temperate climate and the use of these beasts of burden in Europe should probably be looked on as an experiment perhaps by Roman troops who had been previously stationed in Asia (Fig. 12.8).

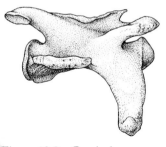

Figure 12.8   Cervical vertebra of a camel

[*] Translated by Rawlinson (1964, I, p. 41).

13 *Reindeer*

ORDER ARTIODACTYLA, FAMILY CERVIDAE, *Rangifer tarandus*

In more temperate regions, men are indebted to the unbounded
liberality of nature for a great variety of valuable creatures to serve,
to nourish and to cloath [sic] them. To the poor Laplander, the
Rein-deer alone supplies the place of the Horse, the Cow, the Sheep,
the Goat, etc.: and from it he derives the only comforts that tend
to soften the severity of his situation in that most inhospitable
climate.   (Bewick, 1790)

The reindeer, or caribou as it is called in North America, is the
sole representative of the genus *Rangifer* and it differs from other
deer in a number of characters (Fig. 13.1, p. 122). Notably, both
sexes carry antlers, the young are born without spots, and the feet
are wide and splayed. The two hooves are unusually deeply
divided and the accessory metapodial bones (Fig. 13.2), which are
absent or vestigial in many artiodactyls, are well developed and
carry small independent hooves, as found also in the elk, *Alces
alces*, and the roe deer, *Capreolus capreolus*. The coat of the
reindeer consists of very thick coarse hairs that are hollow to
increase insulation, and it is very variable in colour especially
amongst domesticated animals which may range from very dark
to pure white (Fig. 13.3, p. 122). The nose is covered with thick
long hair which enables the animal to feed in the snow, and
protects it from the cold.

Old bulls shed their antlers at the end of the year, the young
animals in early spring, and the cows in early summer just before
the calves are born. The rut takes place during the early autumn
around the beginning of October. When they are feeding, reindeer
make a continuous low barking noise and whilst on the move
the accessory hooves click loudly in a characteristic manner that is
assumed to function as a means of keeping individuals in contact
with each other.

In North America caribou are generally larger and have longer
and heavier antlers than the reindeer of Eurasia. On both
continents there are a number of geographical races, as in any
wide-ranging mammal, and these have been described under a
large number of subspecific names. They need not be described
here but on both continents the deer can be divided into an

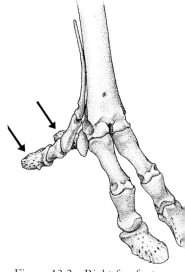

Figure 13.2   Right forefoot
bones of reindeer to show the
lateral digits, arrowed.

open tundra form and a woodland form. Woodland reindeer have less widespread horns than those from the tundra but tend to be larger animals, probably because their level of nutrition is higher.

In Eurasia the tundra reindeer is replaced south of approximately latitude 65° N by the woodland reindeer which lives between latitudes 55° N and 65° N. Wild reindeer are now only found in Norway and the north of Russia, those in Sweden and Finland having been finally exterminated or interbred with domesticated animals at the beginning of the present century. Domesticated reindeer are found in the north of Scandinavia and throughout northern Russia. The caribou was never domesticated, as far as is known, although the European domesticated reindeer has within this century been successfully introduced to North America and to the subantarctic island of South Georgia.

Like the camelids and the equids the reindeer probably evolved in North America, its fossil remains being found in Europe only in the Upper Pleistocene. Also like the camelids, and the asses amongst the equids, the reindeer evolved in a particularly harsh environment and man has exploited their specialized adaptations as part of his own strategy for survival. Throughout the northern hemisphere reindeer inhabit the stark taiga-tundra, and they can bear temperatures of − 50° C, with biting winds and snowstorms a daily occurrence for perhaps nine months of the year. What the reindeer cannot bear are the insects that assault them in July and August and at this time, in the height of summer, they will migrate to higher altitudes where it is cooler and there are fewer mosquitoes, gadflies, and warble flies.

In order that reindeer (and caribou) may flourish in an arctic environment they have evolved adaptations that enable the species to cope with the following four main contingencies:

Finding food. Unlike most species of deer that are woodland browsers, the barren-ground caribou and the tundra reindeer will live on a variety of foods that includes sedges, mosses, leaves, seaweed, lemmings (when they can catch them), and the urine of animals (including man), as well as the lichens that are usually favoured as their chief source of winter food. In order to obtain enough food the reindeer migrate over enormous distances, but usually following the same routes year by year. It is important that they do not overgraze the land because once this occurs the lichens and sedges can take up to thirty years to become re-instated.

Avoiding predation. Since the Pleistocene evolution of the genus *Rangifer* and the movement of reindeer into Eurasia, this deer has had only two main predators − wolf and man. The caribou of North America evolved in an open biotope and early in their history they were preyed upon by wolves. Concealment from predators was not possible so the deer, like many other ungulates of the open plains, evolved a gregarious herd structure and a synchrony of mating and birth activities. As a gregarious mode of life was incompatible with sedentary feeding in an environment where the plants were sparse and very slow-growing the deer became a migrant species that is always on the move. In turn both wolf and man adapted themselves to following their prey.

Everything in the behaviour and physiology of the reindeer is concentrated on thrift and the conservation of bodily resources. Accordingly, female reindeer are usually only mated once by one male and the amount of semen ejaculated is small (Zhigunov, 1968). The calving season is very short, 90 per cent of the calves being born within two weeks during spring, and the calves are able to follow their dams a few hours after birth. Thus the herd structure is held together and the deer are able to keep moving at all times; moreover, as many of the does come into oestrus together over a seven day period it would not be possible for a dominant stag to inseminate many females and fight off rival males at the same time. In this way the wasteful process of strife between rutting males is avoided and the assertion of dominance is much less marked than in other species of deer.

Weather. During the winter, reindeer and caribou will move towards greater tree cover whenever possible but their movements depend on local conditions. In general the deer will move south with the onset of winter and will try to find protection from gales on the edge of forests. Wind and snow are the two great adversities that the deer must survive not only because of the extremes of cold but also because they prevent the animals finding food. In the forests the deer are able to browse off lichens hanging from the trees although the snow cover may actually be deeper amongst the trees than on the windswept taiga.

Insects. Once the winter snows begin to melt, in North America, the female caribou move to the calving grounds which remain the same localities for many years in succession. They are mostly on high windswept terrain and the deer will remain on these lands until midsummer when the plagues of insects begin. Then the deer move away either to higher cooler ground or into wooded country where the mosquitoes are not so numerous. Zhigunov (1968) states that during a mass invasion of mosquitoes each reindeer will lose about 125 g of blood a day and as the arrival of the insects coincides with the moulting period of the deer the animals are particularly vulnerable and will move great distances to escape their attentions.

We can see from this very abbreviated summary of what is known about the behaviour of caribou and reindeer that these deer differ from other species of the Cervidae in two ways that are crucial for their inter-relationship with man: firstly they are highly gregarious and secondly they are not territorial. The reindeer is the only species of deer to have been domesticated (although experiments are now being carried out with the red deer as will be described in a following chapter), and it is the combination of their propensity to travel in large herds, together with their tolerance of intruders, that has enabled man to impose his will on them. In Lapland and in Russia the reindeer fulfill all the roles that in warmer climates are undertaken by cattle. They are ridden, driven, used as pack animals, milked, and eaten. Besides this their hair, hides, hooves, and bones are used for making artefacts and clothing and the rich marrow from the bones is the only common source of fat available for food and lighting.

In the Upper Palaeolithic period of northern Europe before the end of the last glaciation the archaeological evidence indicates that some groups of human hunters depended on the reindeer as much as the Laplanders of today and they lived by following the vast herds of migrating deer. Sturdy (1972, 1975) has reviewed the information available for the Late Glacial reindeer economies in Europe and he concludes that the hunters would have migrated with the reindeer between the winter ranges of the European lowlands as far east as the shores of the Black Sea, and the summer ranges in the highlands of northern and central Europe and Russia. Sturdy postulates that no part of Europe except the high mountains would have been an impossible biotope for the reindeer in the Ice Age winter and the numbers of deer were most probably very high indeed. In these circumstances it is likely that the deer tolerated the presence of man as they will the wolf at the present day, and their flight distance was short. It would therefore have been relatively easy for small bands of human hunters to follow the herds and live off them. There is, however, no evidence for when the deer were first domesticated and although some authors have presumed a very early date for man's first control of reindeer it seems more likely to me that their exploitation was by herding and hunting rather than by true domestication.

The human hunters may have helped to provide the deer with food by scraping away snow and by giving them salt and human urine to lick. Baskin (1974) has described how wild reindeer can be familiarized with human herders to the point where they can be driven into a corral and he remarks that there is such a close resemblance between the herding methods employed by different peoples in central and eastern Asia today that he believes they have a common origin dating back to the time when hunters followed the herds of big game. There is clearly only a hair's breadth distinction between one group of people who follow a herd of animals, killing them when necessary for food and other resources, and another group who persuade the herd to travel at their own direction. According to Baskin this can be done with wild animals simply by shouting, throwing stones, and waving sticks at them, that is by making use of the animals' response to fright which is to bunch together and follow a leading animal.

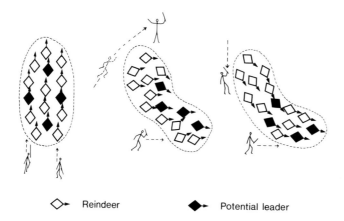

◇➤ Reindeer   ◆➤ Potential leader

This is summarized in Figure 13.4. Baskin also postulates that early man used wolves to keep herds in tight bunches and direct them wherever required. As compact herds of reindeer would fast damage the fragile flora of the winter landscape if they were not allowed to move around he believes that herding and nomadism developed slowly out of hunting.

The bones of domesticated reindeer do not differ from those of the wild form so it is impossible to decide when reindeer were first bred in captivity and therefore truly domesticated. On the other hand the antlers of male deer that have been castrated do differ from those of entire males. Therefore the finding of antlers of a large number of castrates on an archaeological site would indicate that man was controlling the herd and exercising artificial selection in their breeding (the finding of single antlers of this type would be more likely to represent an aberrant wild animal with hormonal imbalance). To my knowledge no collection of the antlers of castrates has been retrieved from a prehistoric site, and all other evidence for domestication such as the harness for riding reindeer and pictorial representations of sledges show that this equipment has been derived from that used for cattle and horses (Fig. 13.5). There is therefore no positive evidence for the domestication of reindeer in the early prehistoric period and although, as with the elephant, man is likely to have been closely associated with these deer since the earliest times they may not have been regularly tamed. The earliest proof for the riding of reindeer comes from the Chinese literature of the first millennium AD (Epstein, 1969).

Figure 13.4 Directing a herd of wild reindeer or other ungulates by means of shouting, waving and throwing stones etc. (after Baskin, 1974).

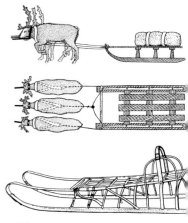

Figure 13.5 Modern sledges and harness for use with reindeer (after Zhigunov, 1968).

14 Asiatic cattle, excluding the zebu

ORDER ARTIODACTYLA, FAMILY BOVIDAE, TRIBE BOVINI

Besides the common domestic cattle of Europe, *Bos taurus*, and the humped cattle or zebu of Africa and Asia, *Bos indicus*, described in Chapter 6, there are four other groups of domesticated cattle that have been of cultural and economic importance in Asia since ancient times. So that they could be viewed in their true taxonomic position a brief outline of all the species within the tribe Bovini was given on p. 65. The Asiatic cattle will now be described in this section in greater detail because their interaction with man is considered to be of a rather different order than that of common cattle and their physical appearance has not been greatly altered by artificial selection.

The domesticated cattle of Asia are the mithan, the Bali cattle, the yak and the water buffalo, and perhaps surprisingly the progenitors of them all still cling to a tenuous existence in the wild. These wild species are the gaur, the banteng, the wild yak, and the wild buffalo respectively. In addition there are four species of wild cattle in south east Asia that have never been domesticated, the little known kouprey from Cambodia and three species of dwarf buffalo or anoa from the Celebes (see p. 66). The survival of these wild species of large bovid is probably due to the fact that until this century their natural habitats have not been greatly under threat. Now, however, they are all either vulnerable or endangered species and will become extinct in the near future unless forceful government action is taken to preserve them.

The wild bovids of Asia fall into three taxonomic groups; firstly there is the gaur, banteng, and kouprey which are anatomically distinct enough from *Bos* to be placed by some taxonomists in the separate subgenus *Bibos*. Secondly there is the yak which again is often placed in a separate subgenus, *Poephagus*, and thirdly there is the water buffalo and anoa which are more distantly related to *Bos*. The water buffalo is classified in the genus *Bubalus* whilst the anoa is either included in this genus or treated on its own in the separate genus *Anoa*.

The wild gaur, *Bos gaurus*, is the largest of all wild Bovidae living today and a bull may have a shoulder height of over two

135

metres. The gaur used to live in high forest throughout peninsular India and Nepal and through to south east Asia as far south as Malaysia, but it was not found on the islands where it was replaced by the wild banteng, *Bos javanicus*, which is closely related to the gaur but is smaller. Both species are now vulnerable; the gaur is only found in scattered remnant herds and the banteng is restricted to a few herds in south east Asia, and on Borneo and Java.

The skull of the gaur and the banteng is markedly different from that of true cattle within the genus *Bos* in having a highly ridged arch between the horns and a concave, or dished, profile to the forehead. The gaur and the banteng can also be distinguished by the hump over the shoulders which is not a fatty lump as it is in the zebu but a crest formed over the extra long neural spines of the thoracic vertebrae. Both the gaur and the banteng and their domesticated forms are dark coloured, black, or reddish-black, and they have strikingly white lower limbs. They also have a white tuft to the rather short tail and the banteng and the Bali cattle have a white rump patch. The kouprey also has this distinctive dark colouring with white feet.

Although the origins of the domestic mithan and the Bali cattle are not proven there can be little reason for denying that they are descended from the gaur and the banteng respectively. Not only do the ranges of the wild and the domestic forms overlap but the conformation and colouring of the cattle are very similar, and their behaviour has many features in common. There is, however, the possibility that domestic cattle of the genus *Bos* have contributed to the ancestry of the mithan and the Bali cattle. In this context it is helpful to consider the patterns of fertility and sterility that are found in the offspring of interbreeding between different members of the tribe Bovini. This may be summarized as follows: the only crosses that are known to be fully fertile are between common cattle and zebu (*Bos taurus* × *Bos indicus*) and between the European and the American bison (*Bison bonasus* × *Bison bison*). With all other crosses, for example between zebu and yak or zebu and gaur, the female offspring are fertile and the male are sterile. No offspring have been recorded from matings between the water buffalo, *Bubalus bubalis*, and any other bovid. Although the gaur and the mithan are said to interbreed and in the past it was a common event for wild gaur bulls to mate with mithan cows, unfortunately there is no documentary evidence for the fertility of the offspring. Similarly there are no records of interbreeding between the mithan and the wild banteng or between the Bali cattle and the mithan or the banteng. It is therefore impossible at present to assess the relationships and origin of these cattle except on the basis of their outward appearance and behaviour.

We are here concerned only with the relationships of the domesticated cattle with man and so no more will be said about

Figure 14.1   Mithan (or gayal), *Bos frontalis*

Figure 14.2   Gaur, *Bos gaurus*

the wild species. The mithan, Bali cattle, yak, and water buffalo will be discussed under separate headings:

THE MITHAN   *Bos (bibos) frontalis*

The history and present-day status of the mithan, or gayal as it used to be more commonly called, has been intensively studied by Simoons (1968) whose fascinating book is a fund of information on every aspect of this rare domestic bovid (Fig. 14.1). Like the gaur the mithan is a woodland browser found at heights of between 600 and 1500 m; it differs from the wild animal in being smaller, much gentler in character, and it has differently shaped horns that are less strongly curved; also the facial profile is flatter and less concave than in the gaur (Fig. 14.2).

The mithan holds a unique place amongst the exploited mammals for its principal value, even today, is as a sacrificial animal. The mithan is only domesticated under the broadest meaning of the term and it is the perfect example of an exploited captive. The cattle are allowed to live free-ranging in the woods, some returning to the villages for protection from predators at night, others remaining outside the control of man except when they are required for sacrifice or barter. As described on p. 67 the cattle are controlled and kept near human habitations by the provision of salt for which they have an insatiable craving.

There is no evidence at all for the ancient history of the domestication of the mithan but Simoons believes that it could well date to the Indus Valley civilizations of 2500 BC, although at the present day mithan are only kept by the peoples of the Assam hills. He makes the hypothesis that if salt were offered to wild gaur only the least nervous and 'wild' individuals would be drawn to it and hence to an association with man. The gentler calves of these animals would again be attracted to man whilst the more aggressive bulls would return to the forest. In this way there would be a selection for tame individuals that would remain near the villages and since the most impressive mithan bulls are most frequently sacrificed the larger individuals would be killed first and the smaller ones would remain as breeding stock. In this way and without any direct interference by man a population of wild gaur could slowly be transformed into a herd of small, gentle mithan.

Simoons relates that mithan are sacrificed by the hill-peoples of Assam on all sorts of occasions, at weddings and burials, at times of misfortune or when people are ill, and at Feasts of Merit. These feasts are a means of raising status and despite their great expense they are performed with regularity by persons who wish to publicly celebrate their success in achieving prosperity. The feasts are highly ritualized, include the sacrifice of many animals, and culminate with the killing of a mithan. The traditional methods of killing the mithan are by strangulation or by chopping, spearing, cutting, or stabbing.

Apart from sacrifice the other reasons that mithan are valued by the hill peoples are for barter, as part of a bride price, and in trade between villages. The mithan are not used as draught or plough animals although hybrids between mithan and common cattle are said to be used as work animals in Bhutan. They are not milked and are not slaughtered for their meat except at the time of a sacrifice when the flesh is cut up and eaten straightaway at the feast. The horns are used as drinking vessels, especially for rice beer, but neither the hide nor the bones are traditionally used economically.

BALI CATTLE *Bos javanicus*

It can be assumed that the typical cattle of Bali, the small island that lies to the east of Java, are the domesticated descendants of the wild banteng that now only live wild in small numbers on Java (Fig. 14.3). The Bali cattle are much more intensively exploited than the mithan and are used as plough animals and as suppliers of meat (Fig. 14.4, p. 139). Besides being found on Bali they are bred in Malaysia, and on Sumatra, Borneo, and Java. Zeuner (1963) reported that they are often interbred with zebu but as with the mithan there seems to be no information on whether the hybrids are fertile or sterile. Also, like the mithan, there is no historical or archaeological evidence to indicate the date at which wild banteng were first tamed.

Figure 14.3   Banteng (wild), *Bos javanicus*

THE YAK *Bos grunniens*

The yak or grunting ox was known by repute to the classical Greeks who called it *poiphagos*, the eater of grass (cf. *Poa* spp.) and indeed the yak is a grazing animal that like the reindeer is accustomed to travel great distances in a harsh environment (Fig. 14.5). These two animals, a bovid and a deer, have many features in common as regards their relationship with man. The domesticated yak differs little from the wild animal except that it is smaller, has shorter and thinner horns, and may be variable in colour. The wild yak lives in desolate mountain areas in the Himalayas up to a height of 6100 m, whilst the domestic form is not found below 2000 m. Yaks look ungainly, partly because they are covered with very long hair and wool, and they have an enormous tail with very long hair coming from its root which is rare in bovids, but they are exceedingly agile climbers and can travel along the narrowest mountain paths.

Figure 14.5   Yak, *Bos grunniens*

Among the Sherpa of Nepal yak herds are kept as a symbol of wealth and prestige and like the Lapps with their reindeer, relatively well-off people, who would disdain the life of a farmer and manual labourer, will undergo extreme hardship in order to tend their herds in the mountains (Epstein, 1977). The yak is an excellent pack and riding animal and can carry loads of up to 150 kg. At high altitudes up to 6000 m an animal can carry a pack or a man at a steady pace for days at a time and still remain in good condition. In some regions the yak is the only

Figure 14.4 Bali cattle (domestic), *Bos javanicus*, see p. 138 (photo I. Glover).

Figure 15.1 Wild rabbits, *Oryctolagus cuniculus*, see p. 146 (photo Geoffrey Kinns).

Figure 14.6 Water buffalo, *Bubalus bubalis*, see p. 140 (photo I. Glover).

Figure 15.2 Wild polecat, *Mustelo putorius*, see p. 148 (photo Geoffrey Kinns).

pack animal available whilst in others it is used as common cattle are, being milked and occasionally slaughtered for meat. As with the reindeer, in adaptation to the cold climate, the scrotum of the male and the udder of the female are small and compact. The udder is covered with hair and the teats are only 2–3 cm long. The milk is golden coloured and has a very high fat content. Yak butter is used in great quantities by the Sherpa as a staple food and also as a lighting fuel.

In all the lower regions where yak are found they are inter-bred with common cattle, either the humpless cattle of Tibet and Mongolia or the zebu. The sires are usually cattle and the dams yaks (Epstein, 1969). The hybrid is commonly known as a dzo. Like the mule the hybrid offspring of cattle and yak surpass their parents in strength and vigour; the females are fertile but the males are sterile. If the female offspring are not back-crossed with either parent, again the female young are fertile but the male are sterile. Back-crossing is not common, however, as the young do not retain hybrid vigour. Hybrids are intermediate in appearance between the parents, having a shorter coat with much less downy undercoat than the yak whilst the females yield larger quantities of milk than the yak cow. The hybrids are preferred for ploughing in Tibet because the yak is said to be too stubborn.

Nothing is known of when the yak was first domesticated but it is probable that there has been a close interaction between man and these cattle ever since the first human immigrations into the high mountains of Asia.

WATER BUFFALO   *Bubalus bubalis*

More than half the peoples of the world depend on rice for their staple diet and it is the water buffalo that enables rice to be cultivated and threshed with the greatest efficiency and economy (Fig. 14.6). Although there is no archaeological evidence to support the supposition, it is likely that the earliest domestication of the water buffalo occurred in the rice growing regions of southern China or Indo-China. Wild water buffalo are rare today and are only found in game parks and in the inaccessible riverine jungles of Nepal and south east Asia. At the present day there are two main groups of domesticated buffaloes, the river breeds of India and the swamp breeds of China and Burma. The swamp buffalo are most like their wild progenitor *Bubalus arnee* in that they usually have straighter more swept-back horns than the Indian breeds. In some of these the horns can be quite tightly curled in slight resemblance to those of the African buffalo, *Syncerus caffer*, to which they are only distantly related. Domesticated buffalo can be white but they are usually slate black with dark skin and only a scant coat of hair. There is a white patch under the chin and there may also be a light patch on the chest. Both sexes have horns which are smaller in diameter and shorter than in the wild species;

Figure 14.7   Sassanid silver dish from Iran depicting the hunting of wild water buffalo (*Bubalus arnee*), amongst other animals, sixth to seventh century AD, Russia (photo Bibliothèque Nationale, Paris).

the size of the body is also smaller in the domesticated breeds. All buffalo depend on living close to water for their well-being.

The river buffalo of India prefer clear running water to muddy conditions and they are better milk producers than the swamp buffalo which are the perfect draught animals for the paddy fields. Cockrill (1967) has described how buffalo are used for ploughing the soil, then when the fields have been flooded they harrow and puddle the mud ready for planting the rice shoots. After planting the buffalo may be rested by turning them out into forest lands to fend for themselves. At harvest time they are reclaimed and used for drawing the sheaves of rice on carts to the threshing floors. Threshing is also carried out by the buffaloes which are made to walk round and round in a tight circle trampling the rice with their feet, in much the same way as is shown for the threshing of wheat in Figure 6.1, where a mule and two cows were used.

Water buffalo are efficient draught animals as long as speed is not required of them and they are allowed to wallow in water during the heat of the day, for they are not tolerant of high temperatures in the absence of water.

Water buffalo, especially the castrated males, are usually tract-able gentle animals that can be managed by anyone with patience, even a small child, and it is likely that their domestication is of very ancient origin. A seal impression from the Mohenjo Daro civilization of the Indus Valley shows that the water buffalo was already domesticated at this period in what is now Pakistan (c. 2500 BC) and another seal impression from a level of similar age was found at the royal cemetery at Ur in Mesopotamia. There is some evidence to indicate that the water buffalo may have occurred wild in the Middle East in the prehistoric period. Bökönyi (1974) has postulated that the finding of a Sassanid silver dish of the sixth to seventh century AD with representations of a hunting scene engraved on it may indicate that buffalo survived in the wild in Iran to this late date (Fig. 14.7).

Neither wild nor domesticated water buffalo were known further west in the Ancient world and they did not reach Egypt or Europe until the early Middle Ages. Water buffalo are restricted to where they can flourish by their need of a hot climate with plenty of water in which they can wallow, and the coarse herbage on which they feed. Buffalo will therefore not survive in northern temperate Europe but they have been common draught and milk animals in Italy and south eastern Europe since about the seventh century AD (Bökönyi, 1974).

During the last 100 years water buffalo have been taken to a great variety of countries where they thrive as work animals in conditions that are hot and swampy. There is a very large number of buffalo now in Brazil and they have been imported as well to New Zealand and the Phillipines.

OPPOSITE
Rabbits from a terracotta moulding in the British Museum (Natural History)

Section III
Small mammals

Small mammals

Of mice and men – it is difficult to pinpoint their relationship. A white mouse living in a child's bedroom is a domestic pet; a brown mouse of the same species living in the larder of the same house is an unwholesome pest to be exterminated as speedily as possible. This is the same situation in miniature as has existed between man, dog, and wolf, for the last 10 000 years, but with the small mammals it seems more difficult to differentiate between friend and foe. No one in the world will eat the common black or brown rat for pleasure but the smoked meat of the giant rats of West Africa, belonging to the same family (Muridae), is relished as a delicacy. In one section of society the red fox will be hunted for sport because it is considered to be a pest, whilst in another it will be bred on fur farms for commercial gain. Our attitudes to the animal world depend on what we hope to get out of it, and with many small mammals there is only a narrow margin between the exploited captive and the troublesome pest that is killed on sight.

Probably the earliest of the small mammals to be domesticated in the Old World was the ferret but nothing is known of where or when this occurred except that it is likely to have been before the Roman period. In Chapter 15 a summary of the natural history of the rabbit and the ferret is given together because, in their relationship with man, they are interdependent. The rabbit and the dormouse were first kept in enclosures by the Romans for food, whilst the Guinea pig was found as a fully domestic animal in South America by the Spanish invaders in the sixteenth century AD. The remaining small mammals that are described in Chapter 16, although they may have been hunted for food and for their pelts for thousands of years, have only been bred in captivity for the last 100 years. They really fall into the category of newly-domesticated mammals that are still in the experimental stage of exploitation. The breeding of fur-bearing mammals in captivity should be much encouraged, however, as the killing of the animals in the wild on a commercial basis in the modern world is inevitably self-defeating. There are just not enough furs to go round.

15 *The rabbit and the ferret*

ORDER LAGOMORPHA, FAMILY LEPORIDAE, *Oryctolagus cuniculus*, RABBIT

ORDER CARNIVORA, FAMILY MUSTELIDAE, *Mustela furo*, FERRET

These two very different mammals are described in the same chapter because the ferret would never have been such a widespread domestic animal if it were not for the rabbit. The wild polecat, the ancestor of the ferret is a specialized hunter of animals that live in burrows, a fact that must have been known to man since ancient times. For like the wolf and the birds of prey that are used in falconry this carnivore was manipulated so that it would hunt its natural prey not for its own food but for the benefit of man – a most cunning arrangement. There is early documented evidence for the use of ferrets to hunt rabbits in the passage by Pliny* written in the first century AD:

There is also a species of hare, in Spain, which is called the rabbit; it is extremely prolific, and produces famine in the Balearic islands, by destroying the harvests . . . It is a well-known fact that the inhabitants of the Balearic islands begged of the late Emperor Augustus the aid of a number of soldiers, to prevent the too rapid increase of these animals. The ferret is greatly esteemed for its skill in catching them. It is thrown into the burrows, with their numerous outlets, which the rabbits form, and from which circumstance they derive their name (*cuniculos*), and as it drives them out they are taken above.   (II, 81)

THE RABBIT

Within the Order Lagomorpha are the pikas, hares, and rabbits. They are an odd group of small mammals that used to be classified as rodents but they differ from them in a number of anatomical characters; notably in that the testes of the male are in front of the penis as in marsupials, and the incisor teeth are peculiarly arranged in having the second upper incisors placed behind (posterior to) the first instead of beside them. Another unusual character of the Lagomorphs is that they have a singular digestive system to deal with the grass and other coarse fibrous vegetation that they feed on: they excrete their faeces twice. The first faecal material consists of moist pellets which the animal eats as soon as they are expelled from the anus and only the second batch of faeces, which differs from the first, is left as

* Translated by Bostock & Riley (1855, II, p. 349).

droppings. In this way the food passes through the gut twice and is doubly digested.

Pikas living today all belong to the genus *Ochotona*. They are small animals that look rather like Guinea pigs but with larger, rounded ears. The genus contains about 14 species, all living in the mountains of Central Asia, with the exception of *Ochotona alpina* which spreads into North America. Pikas have never been domesticated. Until perhaps as late as the eighteenth century AD there was another genus that inhabited the Mediterranean islands, especially Corsica and Sardinia. This pika, named *Prolagus sardus*, is known from written descriptions and from its skeletal remains, but it is not known why it became extinct. Perhaps the *'dasypus'* of Pliny was this animal.

The hares are in nature the most successful group of the Lagomorphs. There are several genera and many species ranging throughout Europe, Asia, Africa, and North America. Hares resemble the rabbit more closely than they do the pikas but they differ in being more solitary in their behaviour and in living above ground. They do not burrow like the rabbit, and no species has been domesticated although the Romans kept them enclosed in warrens or *leporario*. These warrens were also used for keeping captive snails, bees, dormice, and large mammals such as deer.

The European rabbit, which now has a ubiquitous distribution, is the sole representative of the genus *Oryctolagus*. From the end of the Pleistocene until the Roman period it was restricted to the Iberian peninsula, southern France, and possibly North Africa, and if it had not been introduced into the rest of Europe by man it could be now extinct, like the European pika. In fact the rabbit is the best example of that group of mammals, which includes the horse, whose distribution and success as species has been enormously enlarged by the activities of man (Fig. 15.1, p. 139).

Although the Romans were responsible for the spread of the rabbit out of Spain they did not attempt to produce artificial breeds from a domesticated stock, they merely fattened them for food or left the rabbits to live as they would in the wild, except that they were kept in enclosed hutches, areas, or warrens. The extremely fast rate of reproduction of the rabbits, combined with their burrowing habits, enabled them to escape from the Roman warrens and rapidly invade the countryside, but they did not reach the British Isles until the Norman period.

The earliest account of the rabbit in Italy is in the instructions of Varro in the book on agriculture that he wrote in about 36 BC, when he was aged 80, for his wife Fundania who had bought a farm. Talking about hares and rabbits Varro* states:

There is a recent general practice of fattening these, too, by taking them from the warren and shutting them up in hutches and fattening them in an enclosed space. There are then, some three species of these: one, this Italian species of ours, with short fore-legs and long

* Translated by Hooper & Ash (1967, p. 491).

hind legs, the upper part of the body dark, belly white, and ears long [common brown hare]. . . . Belonging to the second species is the hare which is born in Gaul near the Alps, which usually differs in the fact that it is entirely white [mountain hare]; these are not often brought to Rome. To the third species belongs the one which is native to Spain – like our hare in some respects, but with short legs – which is called cony [rabbit]. The conies are so named from the fact that they have a way of making in the fields tunnels (*cuniculos*) in which to hide. You should have all these three species in your warren if you can. You surely have two species anyway, I suppose, as you were in Spain for so many years that I imagine the conies followed you all the way from there.   (III, xii)

Selective breeding of captive rabbits was probably first carried out by Medieval monks who, like the Romans, relished them as food and also ate the unborn and newly-born young. These were not considered to be 'meat' and so could be eaten during 'fasts'. The rabbits were kept in walled and paved courtyards so that they had to breed above ground thus enabling the young to be easily removed. Domesticated rabbits were probably quite quickly bred to be larger in size than the wild progenitor, and the skull became differently shaped with the facial region enlarged at the expense of the cranium which was reduced in size. Colour variation in the pelt was probably not a very early development because although the rabbit has very soft fur it is not much use for clothing because the hairs are rather quickly shed from the skin, a fact that was remarked on by Pliny. At the present day the very wide variety of coat colours that are found in the rabbit are the result of selection for aesthetic rather than practical reasons.

Their propensity for burrowing has meant that rabbits have always been difficult to keep enclosed and they have escaped to breed in larger numbers wherever they have been taken by man. For this reason in the Middle Ages it was popular to have rabbit colonies on small islands where they could be easily controlled. They were also carried by sailors and let loose on oceanic islands so that they could breed and provide a store of fresh meat that would be readily available for passing ships. It is for these reasons that rabbits are now found all over the world and on almost all islands from Lundy to the Falkland Islands. Similarly rabbits were taken by the early European colonists to Australia, where as everyone knows they became as great a pest as the inhabitants of the Balearic islands found them to be during the time of Varro in the first century BC.

The 'wild' rabbits of Europe (except those in Spain) are strictly speaking feral because they are descended not from an indigenous species but from animals that escaped from captivity, even if this occurred as long ago as the Roman period. On the other hand to call them feral would really be too pedantic because these rabbits are now totally adapted to their environment as a wild species. Moreover their captive ancestors were probably almost identical in appearance with the wild Spanish species. The one character that they may have retained from these captive ancestors is that most European 'wild' rabbits are larger than those in Spain.

THE FERRET

In the words of the ninth edition of the *Encyclopaedia Britannica* the ferret is an animal that although exhibiting considerable tameness, seems incapable of attachment and when not properly fed or when otherwise irritated is apt to give painful evidence of its native ferocity. In a word, it is not a trustworthy pet but is a useful partner in the hunting of rabbits and rats. Traditionally, when enlisted in this way the ferret is muzzled or attached to a long lead and allowed to dive down into the animals' burrow or warren. Nets are placed over the outlets to the burrows and as the animals rush out they are caught, or if nets are not used the animals are killed by blows or guns as they emerge. If the ferret is allowed to go freely into the warren it will kill as many animals as it can and will then feed and sleep until it is hungry again. There has to be total silence above ground while the hunt is on otherwise nothing will induce the rabbits to leave their burrows and if the ferret is unmuzzled they may all be killed below ground.

The ferret is descended from a species of polecat, *Mustela*, but it is not clear which species was first domesticated, nor in what country. The earliest mention of the ferret is by Strabo* (III, 2, 6) who wrote in about AD 20 that ferrets were brought from Libya. This statement became fact by tradition and was repeated by Linnaeus who described the locality of the species *Mustela furo* as 'Africa'. The difficulty about this is that the two contenders for ancestry – the European polecat, *Mustela putorius*, and the Steppe polecat, *Mustela eversmanii* – are not found in North Africa. In their skeletal characters the ferret and these two species of polecat are very similar and there are differing opinions about which are the closest in osteological features.

The wild polecat is a larger animal than the ferret, reaching up to 60 cm in length (Fig. 15.2, p. 139). The coat is blackish-brown, very sleak, with lighter underfur, and yellowish patches on the ears, muzzle, and nose. Ferrets are nearly always bred in the albino form, that is yellowish-white with red eyes, and indeed Linnaeus described *M. furo* as having 'rubicond' eyes. The white colour of the ferret is a definite advantage in that it allows the animal to be easily seen and re-captured when taken out hunting. The wild polecat is a hardy animal that can survive the coldest northern winter and in this it contrasts with the ferret which is said to succumb quickly to the cold and has to be kept in a warm sheltered cage. This is perhaps one small piece of evidence in support of the idea of a southern origin for the domestic form.

The ferret, whatever its specific origin, is, like the polecat, a weasel-like animal belonging to the family Mustelidae. These are

* Translated by Jones (1969, II, p. 35).

all solitary carnivores and it would not be expected that they could be tamed in the way that can be achieved with a social carnivore such as the dog. They always remain erratic and recalcitrant in their behaviour.

16 Rodents and carnivores exploited for food and fur

Hominids have always eaten rodents when they could catch them and so will many of the larger primates, such as the baboon, at the present day. Beavers, squirrels, rats, mice, cane rats, dormice, and Guinea pigs, amongst many other rodents have played a large part in the economic and cultural history of man. Only two of these, however – the dormouse and the Guinea pig – were domesticated in the ancient world, the first by the Romans and the second, probably, by the early civilizations of South America. Many rodents and also many carnivores have also been killed to provide furs for clothing (Fig. 16.1), bedding, and floor coverings, and within more recent times a larger number of species have been bred in captivity for this purpose, so that fur-farming has become of considerable economic importance in some countries. It is not intended to give here a comprehensive review of all the rodents and carnivores that have been exploited but to discuss briefly, in a historical context, the interaction of man with some of the more important species.

THE GUINEA PIG

ORDER RODENTIA, FAMILY CAVIIDAE

Cavia porcellus. Guinea pigs do not come from Guinea in West Africa but from South America. They are a domesticated form of a wild cavy, belonging to the genus *Cavia*, but it is not known from which species they are derived. There are various explanations for the incongruous name of Guinea pig; probably they were first called pigs because of their fat little bodies and insistent squeals. The 'Guinea' may have arisen as a misnomer for Guyana on the north coast of South America, or perhaps less likely, because ships carrying Guinea pigs for meat called at ports on the West African coast before proceeding to Britain.

The family Caviidae is confined to the continent of South America and includes five living genera and about 23 species. The genus *Cavia* is found from Colombia and Venezuela southwards to Brazil, Peru, and the Argentine Republic. There are five wild species at the present day: *C. aperea* and *C. fulgida* found in Brazil; *C. nana* and *C. pamporum* in the Argentine Republic; and *C. tschudii* in Peru. Although it is generally

Figure 16.1   The most luxurious clothing is still made from furs. This coat is 'silver fox', a variety of the common red fox, *Vulpes vulpes*, that is specially bred for its fur (photo BM(NH)).

believed that the cavy was first domesticated as a food animal by the Inca civilisation in Peru, or by earlier peoples in the same region, it is *C. aperea* rather than the Peruvian species, *C. tschudii*, that has until recently been considered to be the ancestral species. This is perhaps because *C. aperea* often has a typical crest of upstanding hair on the nape of the neck as is common with many Guinea pigs but there is no other reason why this species should be preferred.

The wild cavies are grey or brown in colour with a pelt that has agouti-type banded hairs (see page 108). They are small round-bodied rodents with small ears and no tail. Although one name for them is the 'restless cavy' they are actually rather calm animals apart from their habit of squeaking a lot.

At the time of the Spanish invasions of South America in the sixteenth century AD the Guinea pig was a fully domestic animal with many colour variations; it was bred by the Incas as a food animal and also for religious ceremonies. Their bodies were often mummified and there are also pre-Inca ceramic representations of cavies. At the present day domestic Guinea pigs are allowed to scavenge freely around the huts of the native Indians and presumably they are killed for food when required. A review of the history of the Guinea pig and its relationship with the wild species has been given by Weir (1974) who makes it clear that much work still remains to be done on the biology, behaviour, and history of this rodent which has now become a very popular pet and laboratory animal.

THE FAT DORMOUSE
ORDER RODENTIA, FAMILY GLIRIDAE

Glis glis. There are six living genera of dormice in Europe and Asia, and one in Africa. The fat dormouse, *Glis glis*, occurs as a wild animal today in woodlands throughout continental Europe and western Asia (Fig. 16.2). It has been introduced within recent times into eastern England. The dormouse was never a truly

Figure 16.2 Fat or edible dormouse, *Glis glis* (photo Geoffrey Kinns).

domestic animal but was kept in captivity by the Romans and eaten as a delicacy. Dormice are nocturnal arboreal rodents slightly resembling squirrels with soft fur and a rather long tail which is said to break and drop off if the animal is caught by it. They are not as prolific as many rodents and only breed once a year. Dormice can be quite aggressive and will jump up and down in one spot, 'purring' and gnashing their teeth when annoyed. Dormice hibernate throughout the winter and prepare for their long sleep by building a nest, storing food, and becoming exceedingly fat beforehand.

Varro* (116–26 BC) wrote precise instructions in his book on agriculture on how to keep and breed dormice (Fig. 16.3):

Figure 16.3

The place for dormice is built on a different plan, as the ground is surrounded not by water but by a wall, which is covered on the inside with smooth stone or plaster over the whole surface, so that they cannot creep out of it. In this place there should be small nut-bearing trees; when they are not bearing, acorns and chestnuts should be thrown inside the walls for them to glut themselves with. They should have rather roomy caves built for them in which they can bring forth their young; and the supply of water should be small, as they do not use much of it, but prefer a dry place. They are fattened in jars, which many people keep even inside the villa. The potters make these jars in a very different form from other jars, as they run channels along the sides and make a hollow for holding the food. In such a jar acorns, walnuts, or chestnuts are placed; and when a cover is placed over the jars they grow fat in the dark.

(III, xv)

The fashion for eating fat dormice was not, however, very long-lasting for, if Pliny is to be believed, by the time that M. Scaurus was consul of Rome in 14 BC the dormouse, along with shellfish and birds, had been banished, at least temporarily from the tables of the rich citizens. It is not clear why such a law was made but it seems to have been because the senators believed that the eating of these small carefully-nurtured animals showed a degree of ostentatious luxury that was not supportable.

RATS AND MICE

ORDER RODENTIA, FAMILY MURIDAE

The Muridae are the dominant rodents of the Far East, there being a large number of genera and species widespread over south eastern Asia and to a lesser extent in Africa. Only two genera, *Micromys* which includes the harvest mouse, and *Apodemus* which includes the wood mouse, are truly indigenous in western Europe. Two species of rat, the black and the brown, and one species of mouse live as commensals of man and therefore have an almost worldwide distribution.

THE BLACK RAT

Rattus rattus. The black rat is a native of Asia Minor and the Orient, and as a carrier of diseases it is believed to have caused a greater number of deaths in the human species than any natural

* Translated by Hooper & Ash (1967, p. 497–8).

catastrophe or war. The most serious of these diseases are bubonic and pneumonic plague, caused by the bacterium *Yersinia pestis* which is carried by rat fleas. Black rats that are infected with plague are bitten by fleas that become 'blocked' by the multiplying bacteria and they will then bite any host they can find, including man, to whom bubonic plague is then transmitted. Pneumonic plague is a highly infectious variant caused by bacteria invading the lungs of a victim so that they are then transferred in the breath or sputum. A comprehensive review of the history of plague and the black rat has been given by Twigg (1978).

Besides bubonic plague rats transmit other serious diseases such as typhus, food poisoning, and rabies. Black rats probably spread through southern Europe on ships from western Asia, probably with the Romans although it is not an animal that is mentioned by Pliny or earlier authors and it did not reach Britain until the post-Roman period.

THE BROWN RAT

Rattus norvegicus. The brown rat is a rather larger animal than the black rat and it is better at burrowing, whilst the latter is more of a climber which is perhaps why it was such a successful inhabitant of ships (Fig. 16.4). The brown rat spread across Europe in the mid-sixteenth century AD and reached Britain in about AD 1720. Its natural environment was probably along stream banks in eastern Asia. As an immigrant to Europe the brown rat

Figure 16.4   Brown rat, *Rattus norvegicus* (photo Geoffrey Kinns).

has been even more successful than the black rat and after its establishment the black rat was ousted from its dominant position in the cities and was driven back to the ports and onto ships.

It is difficult to imagine a more successful mammal than the brown rat, it will live anywhere and eat anything. It is prolific, adaptable, and cunning. Like the black rat it prefers a warm home and it is for this reason that in cold countries rats' are not found far away from human habitations.

Within recent times the brown rat has been domesticated and has become an affectionate pet and a highly successful laboratory animal that is bred in vast numbers for biological and medical research. The domestic rat is usually albino-white or parti-coloured.

THE HOUSE MOUSE

Mus musculus. At the present day the house mouse has an almost worldwide distribution, for, whether man likes it or not, it lives with him (Fig. 16.5). It is difficult to tell where the species was originally indigenous as a wild animal but it may have been in southern Europe and Asia. The house mouse spread north and west with man at an earlier date than the rats, and in Britain the first well-dated evidence for its presence comes (at present) from the Iron Age site of Gussage All Saints in Dorset. The house mouse was well known to the Romans and Pliny wrote on how a visit from a white mouse would bring good luck, whilst a 'singing' mouse 'will interrupt the auspices'. (The singing of mice is well authenticated and sounds like the voice of a weak canary; a single song lasting sometimes for as long as ten minutes.)

There are a number of different races of *Mus musculus*. There is a 'wild form' living out of doors throughout southern Europe, North Africa, and through southern Russia to Mongolia and Manchuria. This form is lighter in colour and has a shorter tail than the northern 'commensal form' that lives closer to man. There is also an eastern 'commensal form' which is found in Japan and Korea and has a dark pelage and long tail.

Despite the unpleasant smell of the male, the house mouse has been tamed and kept as a domestic animal for at least the last three hundred years, probably a lot longer than tamed rats. White, grey, and dark varieties were known in the mid-seventeenth century AD. Today the house mouse is a most valuable laboratory animal which may have a more varied origin than the European pet for populations.have been produced by breeding from both the European and the Japanese forms.

GIANT RATS AND CANE RATS

The many species of giant rats (Fig. 16.6) that inhabit Africa and south east Asia, as well as the giant cane rats of Africa are hunted by man for food, and within recent years there has been a scheme in West Africa to domesticate the giant rat, *Cricetomys gambianus*

Figure 16.5   House mouse, *Mus musculus*

(Family Cricetidae). Ajayi (1975) in giving a description of this project states that in Accra, during one year, 73 tons of meat was for sale in the local market, provided solely from the giant rats known as grass cutters, *Thryonomys swinderianus* (Family Thryonomyidae). This amount of 'bush meat' represented some 15 564 wild animals. The giant rats are usually eaten as smoked meat and are much in demand.

Figure 16.6   Giant rat, *Cricetomys gambianus*, left, Cane rat, *Thryonomys swinderianus*

THE GOLDEN HAMSTER
ORDER RODENTIA, FAMILY CRICETIDAE

Mesocricetus auratus. The domesticated golden hamster that is now such a common household pet and laboratory animal was unknown until 1930 (Fig. 16.7). It may be said that this rodent is the most successful newly domesticated animal, for the entire world population of golden hamsters is descended from one female, unearthed with 12 young, at Aleppo in Syria, and taken from there to Israel. The golden hamster is classified in the large subfamily Cricetinae that, amongst other rodents, includes about 13 more species of hamster, none of which has been domesticated. Hamsters are in general solitary, aggressive rodents and the golden hamster is anomalous in being so docile and easy to handle, but even so it is content to live alone. Wild hamsters range across central Europe and Asia whilst at the present day the wild golden hamster and its near relatives the Ciscaucasian hamster, *M. raddei*, and the Rumanian hamster, *M. newtoni*, are found in western Asia and eastern Europe.

Figure 16.7   Golden hamster, *Mesocricetus auratus*

PROVIDERS OF FURS AND HIDES

Almost all animals that have been killed for food have, throughout human history, also been used to provide skins for clothing and leather for artefacts and shoes. In addition a large number of mammals, particularly carnivores, have been hunted for their pelts alone, the most popular being seals, bears, wolves, and foxes. Ten thousand years ago, in the Pre-pottery Neolithic periods of Jericho, the fox was one of the most commonly killed animals. Its bones were amongst the most frequently retrieved organic remains from the excavations and as they show evidence of chopping and burning the people must have been eating fox meat. It may be assumed that the furs of these foxes were also as much in demand for clothing then as they are now.

Figure 16.8 Chinchilla, *Chinchilla laniger*

Within the last 100 years several species of rodent, in particular the South American chinchilla, *Chinchilla laniger* (Fig. 16.8), and the North American muskrat, *Ondatra zibethicus* (Fig. 16.9), have been bred in captivity for their furs as have several species of carnivores such as the mink, *Mustela vison*, and the several varieties of common fox, *Vulpes vulpes*. As a result of escapes of some of these rodents and carnivores from fur farms into the surrounding countryside there has been a considerable number of introductions of new species into the wild fauna across the continents. Notable amongst these are the muskrat, the coypu (a South American rodent, *Myocaster coypus*, that is known in the fur trade as nutria), and the mink. The mink and the mutant forms of the common fox, for example the cross fox and the silver fox, have been bred in very large numbers in captivity during the last hundred years, as demand for their furs has followed the vagaries of fashion. The farming of silver foxes was begun in 1887 on Prince Edward Island in the Gulf of St. Lawrence, Canada, and by 1940, according to Bachrach (1947), there were 2655 fox fur farms throughout the United States. As the foxes are subjected to careful selective breeding and are reared in captivity purely for economic gain they must count as fully domestic animals, as should the varieties of mink that have been bred in the same way on fur farms since 1886.

The trapping and killing of wild animals for their pelts, apart from being cruel, has a long history of explosive economic profit followed by the extinction, or near extinction, of the species hunted, whether it be the southern elephant seal, the tiger, or the chinchilla. If human beings are to persist with the desire to clothe themselves in the skins of other mammals then the breeding of animals in captivity for this purpose will have to be encouraged and increased. In a world of rapidly expanding human populations and dwindling wild life there is no future for the fur trapper because in the long term neither the hunter nor the hunted can survive.

Figure 16.9 Muskrat, *Ondatra zibethicus*

Illustration on pp. 158–159
European bison, *Bison bonasus*
(photo Geoffrey Kinns).

Section IV

*Exploited ungulates in the
pre-Neolithic Period*

Exploited ungulates in the pre-Neolithic Period

No one would deny that goats, sheep, cattle, pigs, and horses are supremely successful domestic animals and that human populations in most parts of the world have relied on them for some thousands of years, that is in broad terms since the Neolithic period. What archaeologists and biologists do argue about is why, how, and where this domestication first occurred. They also argue about whether, in the earlier periods, the late Palaeolithic and Mesolithic, man hunted his prey in a catch-as-catch-can manner as do other carnivores such as the lion or the wolf, or whether he herded, tamed, followed, or in any other way consciously managed the wild ungulates in his environment.

To mention briefly one such discussion, Jarman & Wilkinson (1972) and Bahn (1978) have raised again, recently, the old contention that some examples of Upper Palaeolithic art show European horses wearing some kind of harness. There has been controversy on this question since Piette (1906) published his belief that the markings on the famous carved head from the cave of St-Michel d'Arudy in the Pyrenees were supposed to represent a rope bridle (Fig. IV.1). The evidence for this bridle is, however, really very insubstantial and it is just as likely that the markings on this carved head, which is less than 5 cm in length, represent the lines of the muscles and lie of the hair in summer coat, as shown in a living horse (Fig. IV.2). The so-called, rope nose-band could equally well demarcate the contrasting light-coloured muzzle that is still characteristic of both the Przewalski horse and the unimproved ponies of today.

No doubt Palaeolithic man did often tame young ungulates and perhaps occasionally he tied them up with primitive ropes and thongs but this is a very different matter from the hypothesis that has been proposed by a number of archaeologists since the time of Piette that animal husbandry was well developed long before the so-called Neolithic revolution. It has been suggested that Palaeolithic man herded reindeer, that Mesolithic man selectively culled and managed red deer and that in western Asia the gazelle was at least semi-domestic in the Pre-pottery Neolithic periods (various authors in Higgs, 1972, 1975). If this were indeed so it is worth speculating on why these animals are

Figure IV.1 Ivory carving of a horse's head, St-Michel d'Arudy, Pyrenees (photo Jean Vertut).

Figure IV.2 Head of a living Przewalski horse in summer coat (photo Geoffrey Kinns).

160

not successful domesticates at the present day and why they were so dramatically ousted during the Neolithic period by sheep and goats. There can be no simple answers to these questions: domestication occurred in different countries at different times, and as a result of complicated interactions between man and animal. Some of these facets have already been discussed, whilst others will be reviewed in the following section on experimental domestication.

Whether or not Palaeolithic and Mesolithic man herded or managed the ungulates that he killed for food there is no doubt that certain species were as important to these early cultures as our domestic livestock are today. The relationships of man with a selection of these ungulates, some of which have remained as hunters' spoil until the present day, are discussed in the following chapter.

17 *The meat supply of hunter-gatherers*

There must always have been people who hunt wild animals and gather wild plants for food in every stratum of every society. Today, small boys will spend all day trying to catch fish in muddy ponds in the suburbs of London, whilst older folk, after a hard day's work in the city will delight in picking blackberries in the public parks. In the modern city jargon these are called 'leisure activities', but in the refuge areas of the world there are still remnant human populations that live entirely by hunting, and gathering wild plants. Examples are the Eskimoes and the Bushmen of the Kalahari desert. These people survive in conditions of hardship that may have been seldom experienced by the peoples who inhabited the world at the end of the last glaciation, but all the same a lot can be learned from them about the way of life of hunting societies, and there is no reason for assuming that their strategies differ markedly from those of prehistoric hunters.

The organic remains that are found in the excavations of early prehistoric sites show that the humans ate a great variety of foods, from the meat of wild cattle, to hedgehogs, foxes, nuts, and seeds. The Mesolithic, in particular, was a period that is described as having an economy based on a broad spectrum of foods, but very often excavation reveals that one or two species of large mammal predominated as the primary source of meat. Throughout Europe these were the red deer and the wild boar, in western Asia the gazelle and in some places the wild goat and onager, whilst in North America it was the bison.

THE RED DEER

FAMILY CERVIDAE

Cervus elaphus. From the huge numbers of bones and teeth of red deer that are excavated it seems that this species supplied man with an abundant supply of meat and raw materials for at least 5000 years in northern Europe whilst in the southern latitudes where the climate was warmer red deer predominated for perhaps 50 000 years. Figure 17.1 shows the results of a survey made by Jarman (1972) of 165 sites of late Palaeolithic and Mesolithic age throughout Europe and it can be seen that red deer remains were

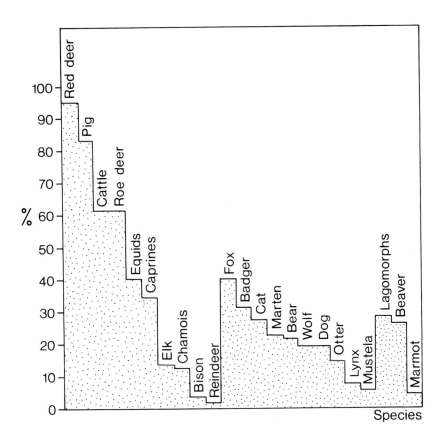

Figure 17.1   The presence of mammals at 165 late Palaeolithic and Mesolithic sites, expressed as percentages (from Jarman, 1972).

found on more than 95 per cent of the sites. It can therefore be accepted that this species was of prime importance to the hunter-gatherers of the early Holocene in Europe; not only did the animals supply meat, hides, bone and sinew, but deer possess a unique physical structure that was of the greatest value, as a raw material to early cultures, this being antler (Fig. 17.2, p. 166).

The males of all the forty or so living species of deer carry antlers, except for the Chinese water deer, *Hydropotes inermis*, and the musk deer, *Moschus moschiferus*. The reindeer is the only species in which the females also regularly have antlers. No other group of mammals possess antlers, which differ from horns in that they consist of solid bone and are shed and regrown every year. Many artiodactyls have horns which consist of a keratinous sheath overlying a bone core that is a permanent outgrowth from the frontal bones of the skull. Antlers, on the other hand, are solid bone without a sheath, and they are cast each year from the pedicles which are short protruberances, again from the frontal bones. Antler is the fastest growing animal tissue in existence and it must be surely one of the most amazing facts of life that the enormous antlers of the extinct giant deer, *Megaloceros giganteus*, were cast and regrown every year of the

stag's adult life (Fig. 17.3). The growth of the antlers of even the European red deer is amazing enough.

In Northern Europe, in the red deer, the old antlers are cast or shed in early spring and new ones begin to grow at once. At first, and until they are fully grown, the antlers are covered with a layer of skin and hair called the velvet, and the bone is well supplied with blood vessels. The antlers are fully grown in about August and then the velvet wears off in long strips and shreds of dead skin. The stags should now be in prime condition and are ready for the rut or mating season. At the present day as soon as the antlers are shed, after the rut, the deer will chew and gnaw at them, actually consuming the greater part of the bone, presumably in order to replace the valuable minerals that they are losing from their bodies. It is not known, however, whether deer needed to eat their antlers in this way in the prehistoric environment where the soil may have contained a much higher mineral content than it does today and the food plants would have been much more abundant. What is known is that the human populations who depended on the deer for their livelihood collected the shed antlers in vast numbers. Maybe they followed the groups of stags and snatched up the antlers as soon as they fell which would not be a difficult task because, as red deer are territorial in their behaviour, they will drop their antlers in more or less the same places each year. On the other hand if the deer did not chew them they could presumably be collected in a more random way whenever they were found lying on the ground.

Antler differs from the limb bones of mammals in that it is solid right through. There is an outer layer of compact bone and this covers a spongy, or trabecular matrix, thereby providing a structure that is extraordinarily strong and yet at the same time very resilient. It is for this reason that antler was one of the most important raw materials of the pre-metal ages. When the crown of an antler of a red deer is removed, it makes one of the most efficient digging tools in the world and although it may not be as long lasting as a metal pick it is quick to make, light to use, and is easily replaced (Fig. 17.4). Another common use for antler, especially on the continent of Europe was for hafting stone axes, as shown in Figure 17.5.

With an abundance of food and the excellent raw materials of wood, bone, flint, and antler it is difficult to see what the Meso-

Figure 17.3   Antlers of the extinct giant deer, *Megaloceros giganteus*, 3 metres across (photo BM(NH)).

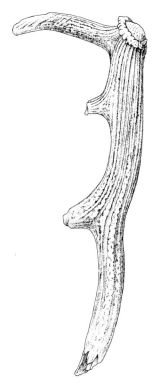

Figure 17.4   Antler pick from the Neolithic flint mines of Grime's Graves, Norfolk, England.

Figure 17.5   Polished stone axe in antler sleeve, Neolithic lake village, Switzerland.

lithic people of Europe lacked. It is therefore worth speculating why this economy was displaced so dramatically by agriculturalists who brought with them imported domesticated sheep and goats, animals that were foreign to Europe and did not survive well in forested areas. What was so special about sheep and goats, which do not even have antlers, and why were the red deer not domesticated and bred to produce docile animals that could be milked?

As explained briefly in Chapter 5 it is a question of the behavioural patterns of the different groups of mammals. It is just not possible to impress upon adult red deer that they are a part of human society which is in effect what has been achieved with all the species of true domestic mammals. As with any animal it is of course possible to keep and breed deer in captivity and the modern methods of overcoming the difficulties of managing red deer are reviewed in outline in the following chapter.

Those who argue, as does Jarman (1972), that the Mesolithic peoples of Europe actively managed the herds of red deer, selectively culling the males and even husbanding the herds, may forget that the numbers of the deer must have enormously outnumbered those of the humans. The forests must have been teeming with wild life and it would require only as much effort as it takes a pack of wolves to kill a wapiti in Canada today, for a group of human hunters to kill a deer. This does not imply that the humans were not thoughtful people, as intelligent as we are today, able to plan ahead, and with traditions and rituals of equal complexity, but in a balanced ecosystem it is unnecessary for a predator, either human or animal, to interfere with the way of life of its prey. In human society this only occurs when the numbers of the prey species become dangerously low and when this happens the dominance hierarchy takes over and the elite forbid the killing of the animals by the common people. This is what occurred, for example, throughout northern Europe in the medieval period when the Laws of Venery were set up and the cruellest punishments were inflicted on those caught poaching. It is very well known how in the time of William the Conqueror anyone caught killing a hart or a hind was blinded by order of the king. But this sort of control requires a highly sophisticated form of government and it would not apply to bands of hunter-gatherers living perhaps rather as the Kalahari Bushmen do today, but of course in a cold climate. Such people would kill whatever they could whenever they could, although they might prefer male animals to females, for the sake of their horns or antlers.

As far as predator-prey relationships were concerned in the human context it seems clear that at the close of the Mesolithic period the ecosystem was no longer balanced. There is still much to learn about the environment at this period but it is known that the climate was changing, the forests were being cleared by man, and the domestic dog was a new aid in the slaughter of large numbers of animals and probably greatly increased their

Figure 18.2 Cheetah hunting Blackbuck from the Akbar nāma. Late sixteenth century AD, see p. 180 (photo Victoria and Albert Museum).

Figure 17.2 Red deer, *Cervus elaphus*. Stag with fully grown antlers, in rut, see p. 163 (photo Geoffrey Kinns).

Figure 18.6 Spotted deer, *Axis axis*, being transported in boats, Andaman Islands, see p. 182 (photo Raghubir Singh).

Figure 18.7 Red deer hind, *Cervus elaphus*, wearing a halter, with keeper, Popielno Biological Station, Poland, see p. 182 (photo G. Barker).

Figure 18.8 A Blackbuck, *Antilope cervicapra*, with his keeper, Indian miniature, see p. 184 (photo Victoria and Albert Museum).

overhunting. The practice of agriculture was beginning to spread and far from the herd management of red deer what became essential was to keep them off the growing crops and drive them into the forests. In northern Europe pigs and cattle, perhaps often domesticated from local stocks of *Sus scrofa* and *Bos primigenius*, may have slightly preceded the imported sheep and goats which came from the Near and Middle East. The use of pottery and the storage of food, both plant and animal, became widespread and human cultural systems were transformed. Red deer could not adapt to this new way of life and so have remained as wild creatures that have been the sport of kings until the present day, when renewed efforts are being made to turn them into 'farm animals'.

MYOTRAGUS BALEARICUS
FAMILY BOVIDAE, SUBFAMILY CAPRINAE, TRIBE RUPICAPRINI

A little known example of the replacement of wild hunted animals by imported domestic livestock has recently been described from the Balearic islands in the Mediterranean. It concerns the extinct goat-like mammal, *Myotragus balearicus*, which was previously believed to have become extinct in the late Pleistocene on Majorca and Minorca, the only islands from which its remains have been identified with certainty (Fig. 17.6). This strange animal has no common name; it is called the 'cave goat' by Kurtén (1968) but it is unlikely that it spent very much of its waking life in caves. Myotragus, as it will be called here, was a bovid that probably belonged to the tribe Rupicaprini within the subfamily Caprinae. If this is so then its nearest European relative would be the chamois, *Rupicapra rupicapra*, that lives on the steep mountain slopes of the Alps and the Pyrenees and eastwards along the mountain chains of Asia.

Figure 17.6   Reconstruction of *Myotragus balearicus* from skeletal evidence.

The excavation of cave sites, in particular that of Muleta by Waldren on Majorca, has revealed that this aberrant, highly specialized caprine survived until the Neolithic period, around 2500 BC, and its remains have actually been found intermingled with those of domestic sheep and goats (Kopper & Waldren 1967). Like many animals that evolve on small islands where there are no predators and an absence of the usual assemblage of competitors for food, myotragus diverged very greatly from its mainland ancestors and it must have been a peculiar animal to look at. It was a small ruminant, weighing about 14 kg, that probably had the appearance of a small, squat goat with short, straight, sharply-pointed horns.

Myotragus, like all other bovids, had no upper incisors, but it was unique in that it had only two lower incisors, instead of six, and these were open rooted and continuously growing. These teeth look rather like the incisors of a beaver, which is a rodent and therefore only very distantly related to the caprines. It is likely that myotragus fed on the same type of woody plant tissue

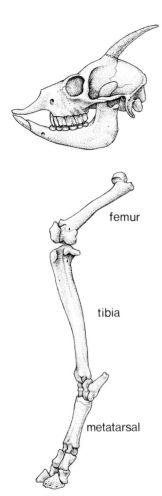

femur

tibia

metatarsal

Figure 17.7   Skull and left hind limb bones of *Myotragus balearicus*

Figure 17.8   Dorcas gazelle, *Gazella dorcas*

as the beaver and it may even have used its incisor teeth for stripping the bark off trees, for grass would have been in short supply on the windy, arid, Mediterranean islands during the Pleistocene, as it is now. Another peculiar feature of myotragus was the unusual shortening of its limb bones (Fig. 17.7); in the forelimb the humerus and metacarpal were remarkably short compared to the radius and ulna, whilst in the hind limb the femur and metatarsal were very short compared to the tibia. Moreover the distal tarsal bones, that is the scaphocuboid and the large cuneiform bones were often fused to the proximal epiphysis of the metatarsal bone. The peculiar proportions of the limb bones were presumably an adaptation that enabled the animals to jump from crag to crag on the steep slopes of the islands. On flat ground they probably moved in a rather slow, stiff-legged kind of way but as there were no predators from which swift flight would be required this did not matter.

When the Neolithic people arrived on the Balearic islands they probably found a population of myotragus that was only controlled by the amount of food available and there may have been very large numbers of these strange-looking 'goats'. They would have provided a ready supply of meat for the humans but at the same time the myotragus would have devoured any grain or plants that were being cultivated. Waldren believes that some attempt was made to corral the animals and even to remove their horns whilst they were alive, that is to poll them for easier management. This may have occurred but it is evident that, in the long term, domestication of myotragus was not successful; this aberrant creature was exterminated and the farmers on the Balearics returned to their ubiquitous sheep and goats.

THE GAZELLES OF WESTERN ASIA
FAMILY BOVIDAE, SUBFAMILY ANTILOPINAE, TRIBE ANTELOPINI
There are three species of gazelle that were common throughout the Near and Middle East until recent times, each inhabiting a particular region or biotope, but also overlapping in their distribution in some areas (Figs. 17.8–10). The bones, horn cores, and teeth of these gazelles are not easy to tell apart when they are found in a fragmentary state on an archaeological site. In general, however, the smallest species is the dorcas gazelle, *Gazella dorcas*, whose preferred habitat is acacia scrub and semi-desert. Next in size is the Arabian gazelle, *Gazella gazella*, which is found in areas with a rather higher rainfall than the dorcas gazelle and is common in the mountains and foothills of the Arabian peninsula, but not in the central deserts. The largest species is the rhim or goitred gazelle, *Gazella subgutturosa*, which is also more heavily built than either of the other two and is distinguished by differently-shaped horns in the male and by the absence of horns in the female. In the females of the dorcas and Arabian gazelles there are small sharply-pointed straight horns. The goitred gazelle

has the most easterly distribution of the three species and ranges across western Asia into Afghanistan and northern Tibet. It does not leap or bound as much as other gazelles but can run exceedingly fast. The preferred habitat of the goitred gazelle is on steppe lands and semi-desert where dwarf shrubs and annuals are in plentiful supply.

Almost all the Upper Palaeolithic and Natufian (as the Mesolithic of the Near East is termed, see chart, Appendix II) sites of western Asia have provided remains of one or other of the species of gazelle, sometimes in very large numbers indeed. Perhaps the best known of these sites are the Wady el-Mughara caves near Mount Carmel, close to the Mediterranean coast of Israel, (Bate, 1937). The two caves of Mugharet-el-Wad and Tabūn, excavated in the 1930s, have revealed a long series of rich deposits that are of great importance for archaeology, biology, and bio-geography. Within the deposits the numbers of gazelle remains fluctuated greatly and seemed to alternate with those of fallow deer, *Dama mesopotamica*. Bate argued from the relative numbers of these species that the climate must have been hot and dry when gazelle predominated and cooler with a higher rainfall when the remains of fallow deer were in the majority. This being because gazelle normally live in arid open-plains country whilst deer are woodland browsers.

Figure 17.9   Arabian gazelle, *Gazella gazella*

Whether or not these deductions were correct, and there has been much controversy on this question in the last twenty years, it is sure that gazelle remains greatly outnumbered those of any other animal in the Natufian levels of the excavations. This pattern can also be seen in the faunal assemblages from a number of other sites in western Asia of the pre-Neolithic period and just as Europeans at this period had an economy based on deer so the people of the Near and Middle East were predominantly hunters of gazelle.

Figure 17.10   Goitred gazelle, *Gazella subgutturosa*

Western Asia is, however, like Europe a very large continental area with a great variety of altitudes and climates so that to claim that one species of mammal predominated over the whole of either region is a facile simplification. Just as people living in the mountains of Europe hunted chamois and ibex, so in the high lands of the Near and Middle East they killed wild goats, sheep, and equids for meat. On the other hand it is true that gazelle remains are extraordinarily common on most sites of Upper Palaeolithic, Natufian, and the earliest phase of the Pre-pottery Neolithic, as shown for example at Jericho (Fig. 5.11). This has led some archaeologists (for example Legge, 1972, 1977) to suggest that the gazelles were, if not domesticated, at least managed and herded in a controlled manner. But here again it may have been forgotten how enormously common herds of hoofed mammals were likely to have been on the grasslands of western Asia before man became fully established as the master predator. When this did occur and meat began to be short, he started to keep a

Figure 17.11   Hand feeding of oryx, *Oryx dammah*, from Tomb 3, Beni Hasan, *c.* 1900 BC (from Newberry, 1893).

walking larder always to hand, that is he enfolded certain species of animal within his own society.

Gazelle, like deer, were not amenable to this manipulation. They can be tamed but they cannot be bunched up together or driven with dogs, nor will they travel with nomadic peoples over hundreds of kilometres as will sheep, goats, and cattle. A crucial difference may be mentioned here between sheep and goats, and reindeer and other migrating herd animals; the former can be driven in the direction that the shepherd wishes to go, but with reindeer and indeed gazelles, man has to follow the herds. He can drive the animals into corrals or for short distances but he cannot alter the course of their migration routes. Mendelssohn (1974) has described, in a quoted translation from a Hebrew writer, the method used by Bedouin for trapping gazelles and this is repeated here because it is likely that this procedure has been in use for as long as human hunters have lived in western Asia:

In order to trap several hundred gazelles at once, the bedouin enclose a large triangular area which extends over many kilometres. In the wall, which is higher than a man, are places which are lower, and before each one a deep trench is dug. When the bedouin saw a migrating flock of gazelles, they drove them from all directions into the broad opening of the corral. The gazelles were not afraid, as the walls, built from desert stone, were similar to their surroundings. When several hundred gazelles had entered the corral, the bedouin closed in on them, running from left to right, shouting ferociously. Then the frightened gazelles tried to escape, jumping over the wall at the lower parts of it, and fell into the trenches outside of them . . . Then they loaded the gazelles onto their camels, brought them to the camp, skinned them and salted the meat.

The preferred habitat of all species of gazelle is grassland and desert scrub where they are the prey of a great variety of predators that include man, wolves, large cats, hyaenas, and even eagles. Their method of escape is by swift flight in response to the quick reactions of their acute senses. Gazelle are therefore intensely nervous animals and will bolt at the slightest fright. If penned they will batter themselves to death against a fence in their frantic efforts to escape. Furthermore, like deer, they are territorial by nature and the males are unlikely to breed well in captivity over a number of generations especially in conditions of primitive enclosure.

Legge (1972) cites the ancient Egyptian pictures of gazelle and other antelopes being hand fed as evidence for their domestication (Fig. 17.11). These pictures, however, only show that the animals were kept as pets, as curiosities, or for sacrifice; if they had been truly domesticated, that is bred in captivity, isolated from the wild species, then they would have been described as such in the classical literature and would probably remain as domestic animals today.

THE AMERICAN BISON
FAMILY BOVIDAE, TRIBE BOVINI

Bison bison. The largest animal that will be described here as a principal resource for prehistoric human hunters is the American bison, *Bison bison* (Fig. 17.12). Although often called 'buffalo' in the United States, this bovid is not closely related either to the African buffalo or to the Indian water buffalo. It is, however, a near relative to, and is now considered to be conspecific with, the European bison, *Bison bonasus*, as the two species will interbreed to produce fertile offspring in both sexes.

From the end of the Pleistocene until recent times bison were the dominant ungulates of the Great Plains of North America and they ranged from Canada to Mexico. Bison were hunted continuously from the time of man's first arrival on the continent

Figure 17.12 American bison, *Bison bison*

but managed to maintain their high numbers until the guns of the nineteenth century Europeans led to their near extermination. After the Spaniards brought domestic horses to America in the sixteenth century AD the Plains Indians killed the bison with bows and arrows from horseback. Before this they trapped them and drove them into ravines or narrow valleys, much as has been described for the way that the Bedouin kill gazelle.

In the remarkable account by Wheat (1967) a step-by-step reconstruction is given of how a band of perhaps 150 Palaeo-Indians drove 190 bison to their death in a gorge, or arroyo, during spring time in about 6500 BC. The site of this kill is in southeastern Colorado just below the northern edge of the Arkansas river valley and it is known as the Olsen-Chubbuck site, after its finders. The deposit of bison bones that filled the bottom of the gorge was about 50 metres in length and the proportions of animals killed were calculated as, 46 adult bulls, 27 immature bulls, 63 adult cows, 38 immature cows, and 16 calves. From the positions of the skeletons, the cuts on the bones, and the stone implements that were found with them the excavators were able to deduce from which direction the herd had been driven into the gorge, from which animals meat had been removed, and how the butchery had been carried out. The animals that fled into the bottom of the gorge first had not been handled by the hunters because they could not be reached, so they had been left to rot along with the rest of the carcasses that were not required.

This site is a classic example of overkill and no doubt played a part, however infinitely tiny, in the gradual progression towards extermination of the bison from the Great Plains. There has been a trend within recent years, amongst archaeologists, to argue that prehistoric man selectively killed his prey and managed the herds of ungulates in his environment so as to preserve an ecological balance. It can equally well be argued that man enjoyed killing as much 10 000 years ago as he did during the last century when travellers would shoot bison from railway trains. If the most efficient way to obtain meat was to drive a whole herd of bison or gazelle into a trap and then to remove as much meat as could be carried back to the camp it is unlikely that the hunters would be remorseful about the surplus killing. Nor is it likely, with a highly successful species such as the bison, that it would be permanently detrimental, although it is known that human hunting has certainly exterminated many species of mammals and birds and is continuing to do so.

We tend to think of the wanton slaughter of animals as something 'unnatural' that would not have been practised by prehistoric hunters who were living in a balanced ecosystem. Kruuk (1972) has, however, studied the question of surplus killing by other carnivores and has shown that it is relatively common when the prey animals fail to make the correct flight response. He cites an example of 19 hyaenas killing 82 Thomson's gazelle in the Serengeti National Park in Tanzania, in one night. The hyaenas had come upon a herd of sleeping gazelle and had quietly walked from one animal to the next, killing or maiming them until they were all dead. Equally, everyone knows how a fox or a wolf will kill a flock of chickens or other farm animals if they can get at them.

The inference is that all carnivores will kill when they are hungry. They will cease to make the effort to hunt when they are satiated with food, but if the prey does not respond when its flight distance is encroached upon then the killing may go on. Human hunters will follow this pattern today and it is really unlikely that in the prehistoric period the world was peopled with noble savages who were consciously aware of the need to conserve their resources. To gain some idea of the tremendous wealth of these resources it is only necessary to read the accounts of European hunters travelling through lands where the gun had been in little use before their arrival.

Harris (1839), for example, described his journey through South Africa when the land was still only principally inhabited by its indigenous peoples and animals. The book makes painful reading for Harris makes such comments as (p. 220):

The country now literally presented the appearance of a menagerie; the host of rhinoceroses in particular, that daily exhibited themselves, almost exceeding belief.

In another passage (p. 224) he writes:

Every open glade abounds with the more common species of game, such as the brindled gnoo, hartebeest, sassayby, and quagga, together with the ostrich and wild hog.

Whilst earlier in the book, this hunter who proudly claims in the first sentence of his preface to have a 'shooting madness', described a plain that seemed alive with quaggas and brindled gnus (p. 71):

The clatter of their hoofs was perfectly astounding, and I could compare it to nothing but to the din of a tremendous charge of cavalry, or to the rushing of a mighty tempest. I could not estimate the accumulated numbers at less than fifteen thousand.

In attempts to reconstruct the environment of the hunter-gatherers of the early Holocene it has become an accepted practice to apply ethnographic parallels from the relics of these societies that survive today; the Eskimoes, the Bushmen, and the Australian Aborigines to name the best known. These comparisons are surely valid and useful, and works such as that of Yellen (1977) should contribute greatly to our understanding of the prehistoric environment, but perhaps the models would be more realistic if the descriptions of early nineteenth century hunters such as Harris were also taken into account.

OPPOSITE
Musk ox, *Ovibos moschatus*
(photo A. J. Sutcliffe)

Section V

Experimental domestication and game ranching, past and present

Experimental domestication and game ranching, past and present

Individuals of almost every species of mammal have been kept in captivity at some time or other, and tamed, but as we have seen only a few species have undergone the full process of domestication. Many affluent societies of the past, as well as at the present day, have enjoyed keeping menageries of exotic animals. The art of Ancient Egypt makes it evident that in the Old Kingdom (2686–2181 BC) almost every local ruminant was kept in captivity and the oryx, gazelle, hartebeest, and addax antelope were all shown being fed by hand and wearing collars round their necks.

Ungulates were not the only mammals to be fed by hand by the Ancient Egyptians for, as is often quoted with some astonishment, hyaenas appear to have been stuffed, like geese are at the present day for pâté de foie gras. They are shown lying on their backs with food being thrust down their throats, but what part of their anatomy was valued after this treatment is not known.

The Ancient Assyrians kept animals in game parks for hunting, much as deer and exotic antelopes are kept on ranches in the New World today, whilst the Romans, as is well known, kept every possible creature for their circuses and menageries. They also kept game parks as described by Columella*:

Wild creatures such as roe deer, chamois and also scimitar-horned oryx, fallow deer, and wild boars sometimes serve to enhance the splendour and pleasure of their owners, and sometimes to bring profit and revenue.
(IX, 1)

Columella explained that wild animals should be kept in enclosures close to the farm so that the sight of the hunt could delight the landowner without his having to travel far, and also when feasts were held and game meat was called for it could be readily available. Pliny also wrote of animals that were,

neither wild nor tame, but of a sort of intermediate nature.

and it is intriguing that he included the dolphin amongst these.

Within recent years there has been a surge of interest in the behaviour and learning ability of the smaller Cetacians, particularly the dolphins, and many are now kept in oceanaria. Dolphins

---

* Translated by Forster & Heffner (1968, p. 421).

are highly social animals and they soon develop relationships with the humans around them that seem to be essentially anthropomorphic. These relationships are not based on the exploitation of the animal, as is inherent in the process of domestication, and they are therefore better described as the socializing of two different species of mammal. A parallel situation can occur between man and apes such as the chimpanzee. Neither the dolphin nor the chimpanzee can be counted as domesticated animals and although the study of their behaviour is obviously of the greatest importance for the understanding of ourselves it falls outside the scope of this book.

The cheetah is the only carnivore to be included in the following chapter, within this section. It is also one of the few mammals that has undergone wide-scale domestication in the past but which is now no longer exploited except by hunters for its pelt.

The ranching of ungulates for the commercial use of their meat, hides, and horns has a number of advantages to both animals and man. It preserves the species from extinction, it allows controlled hunting of people who enjoy this 'sport', and by keeping a variety of species that feed on different kinds of vegetation it is a more effective way of managing the land than by the restrictive grazing of domestic livestock alone. It is finally being realized, and not before it is almost too late, that to exploit the indigenous herds of African ungulates is a more efficient way of obtaining protein from the arid regions than to kill the game and replace it with imported cattle, sheep, and goats. The potential for the ranching and farming of some of these ungulates is discussed in the next chapter.

18 The cheetah, sirenians, deer, and bovids

THE CHEETAH

FAMILY FELIDAE

Acinonyx jubatus. Apart from its spots the cheetah is not very like the other large cats in the family Felidae. It is the swiftest and perhaps one of the most beautiful of the carnivores (Fig. 18.1). The cheetah was formerly widespread throughout India, western Asia, and the savanna country of Africa. It is, however, now a vulnerable species found in small numbers in north, east, and southern Africa, whilst in Asia it is restricted to the arid regions of Afghanistan, Turkestan, and eastern Iran.

The history of man's relationship with the cheetah has been a strange one, and if it were not for the fact that it will only very rarely breed in captivity it is likely that the cheetah would be a common domestic animal today. Both the Ancient Egyptians and the Assyrians kept tame cheetahs and used them for hunting but it was during the time of the Moghul emperors of India that they became immensely popular, that is from the fifteenth to the eighteenth centuries AD. Akbar, the greatest of the Moghul emperors, who lived from AD 1542–1605, is said to have kept 9000 cheetahs during his reign and to have owned 1000 at a time. All these animals were caught in the wild either as adults by placing snares at their scent-marking posts or as young, and it needs little thought to estimate the number of wild cheetahs that were needed to supply the nobility of India on this scale.

Cheetahs are unlike other cats in that they do not pounce on their prey but chase it over a short distance and catch it with great speed. It is this characteristic that was exploited by the Indian nobles. The cheetahs would be taken to the scene of a hunt in a specially made cart or cage with a side that could be let down for their release. Before the hunt the cheetahs would be blind-folded and held with a collar and chain; then when the game were sighted, these being usually gazelle or the Indian blackbuck, the cheetahs would be let loose. They would then stealthily approach the prey and swiftly leap towards it. If the antelope was success-fully killed the cheetah would be given some of its meat or blood to consume but if not, it would return of its own accord to its cart.

Figure 18.1 Cheetah, *Acinonyx jubatus* (photo Geoffrey Kinns).

The cheetahs of the Indian princes were kept in an order of rank according to their success at hunting (Fig. 18.2, p. 166) and Akbar's top cheetah had a drummer to go in procession before it (Mungall, 1978).

Despite the heavy toll on the wild stock, cheetahs remained available to be caught in India until the end of the last century. They finally succumbed to the loss of their grassland habitat as well as to over-capture and hunting for their skins, and as a result the cheetah can no longer be counted as a domesticated animal. It may be noted that the cheetah is one of the most difficult animals to breed in captivity and there was no record of successful births in a zoo until 1960.

THE SIRENIANS
ORDER SIRENIA, FAMILY DUGONGIDAE
Dugong dugon, DUGONG. FAMILY TRICHECHIDAE
Trichechus manatus, NORTH AMERICAN MANATEE
Trichechus inunguis, SOUTH AMERICAN MANATEE
Trichechus senegalensis, WEST AFRICAN MANATEE

The sirenians are an anomalous group of aquatic mammals that live in the shallow coastal waters of the tropics where they have few natural predators apart from man. They are all entirely herbivorous and feed on vast quantities of seaweed and other water plants. The dugong, which is found on the coasts of East Africa, the Red Sea, Indian Ocean, South Asia and Australia, will only live in marine waters but the manatees are more adaptable and will flourish in rivers and lakes. It has been suggested by Bertram & Bertram (1968) that the sirenians should be conserved and exploited in a controlled manner for their meat, and experiments have also been carried out with the introduction of manatees to weed-infested lakes and rivers in order to keep them clear. The same authors (1963) reported that three manatees in Guyana, South America, kept two reservoirs free from weeds for ten years when previously they had to be cleared by hand every few months (Fig. 18.3).

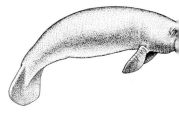

Figure 18.3   South American manatee, *Trichechus inunguis*

As with the cheetah and many other semi-domesticated animals, however, manatees cannot be exploited in the future unless they can be bred successfully in captivity and this has not yet been achieved.

THE DEER
FAMILY CERVIDAE
There can be no living species of deer that has not at one time or another contributed to human economies. The reindeer is familiar as a domesticated animal whilst all the other Eurasian species have been hunted or kept in a semi-domesticated state. The deer that is probably most amenable to domestication, apart from the reindeer, is the elk.

THE ELK
Alces alces. The elk is a forest-living inhabitant of the northern

Figure 18.4   Elk, *Alces alces*

latitudes of Europe, and North America where it is called the moose. Today the elk (or moose) is found in northern Scandinavia, the east of Poland to Siberia, and eastwards to northern America and Canada. Previously, however, this deer was found further south and west and was present in Britain during the Mesolithic period. The elk (Fig. 18.4) was one of the favourite animals of the Scythians and it is much figured in their art.

Zeuner (1963) described how the elk had been tamed and used as a riding animal in Sweden as late as the beginning of the last century. Elks were also milked but there is no evidence to indicate whether these deer were domesticated in antiquity. Elk are, however, proving successful as new domestic animals.

Within recent years elk have been re-domesticated in the Pechero-Ilych National Park at the foothills of the Ural Mountains in northern Russia, where there may be a covering of snow up to one metre deep for seven months of the year (Yazan & Knorre, 1964). In 1949 it was decided to set up an experimental farm for the domestication of the elk which numbered some 3000 in the park. The intention was to produce an animal that was as useful in the taiga zone of Russia as the rendeer is in the tundra and it appears that the project has been most successful. As a source of meat the elk is highly productive, the cows can be readily milked and will return at the appointed time, of their own accord, to the farm for milking. Castrated bulls are also used for draught animals and can be trained either for riding or for pulling a sled.

COMMON FALLOW DEER, *Dama dama*
PERSIAN FALLOW DEER, *Dama mesopotamica*
It is difficult to ascertain the original distribution of the fallow

Figure 18.5   Fallow deer, *Dama dama* (photo Geoffrey Kinns).

deer because the common form has been extensively moved about by man since the Roman period (Fig. 18.5). *Dama dama* was formerly, however, certainly wild around the whole of the Mediterranean region and perhaps also in North Africa, whilst *Dama mesopotamica* occurred in western Asia. The Persian fallow deer differs from the common form in the shape of its antlers which are less palmated and also in its greater size; it is, however, near extinction at the present day. Like most species of deer the fallow flourishes best in a woodland environment.

There is evidence that man took the Persian fallow deer to the island of Cyprus during the Neolithic period for the antlers of this form have been identified from the Neolithic sites of Khirokitia and Erimi (King, 1953).

Although the fallow deer has never been domesticated the species is easily tamed. It was probably first brought to the British Isles by the Romans and has been a common inhabitant of deer parks since Norman times. Figure 18.6, p. 167 shows how easy it is to carry living deer about in boats. This picture is not, however, of fallow deer but of a similar-sized species, the spotted deer of Asia, *Axis axis*, photographed on one of the Andaman Islands in the Indian Ocean.

RED DEER, *Cervus elaphus*

The importance of the red deer to the hunter-gatherers of northern Europe in the early Holocene has been discussed in the previous chapter. Since that period red deer have remained an important species, but until very recently always as a beast of the chase rather than as a domesticated animal (Fig. 18.7, p. 167). Like fallow deer they have been moved around and also they have been kept in deer parks. The antler of red deer was a most important raw material until the present plastic age when its use has become restricted to objects of aesthetic appeal.

Within the last decade the demand for venison has increased and this, in conjunction with the decline of the hill sheep, and a need to manage upland areas in Britain has led to the promotion of deer farms. An experimental farm for red deer was started by the Rowett Research Institute at Aberdeen, Scotland and its progress has been fully documented by Blaxter *et al.* (1974). Young calves of red deer were first of all caught in the wild and hand-reared. These bred freely in conditions of captivity but the progeny cannot really be counted as successful domestic animals. This is because the red deer is a highly territorial species and the stags become most unpredictable in their behaviour, especially at the time of the rut. In order to make them easier to handle the antlers are removed by sawing as soon as they lose their velvet but as can be imagined this is a difficult task and not practical for a large herd. The hinds can also be untrustworthy and aggressive.

Despite the difficulty in management the exploitation of red deer for their meat is highly productive and it is more efficient to keep the animals in a semi-domesticated state than as wild animals that can only be killed by hunters. The traditional deer parks of Britain are probably the best way of achieving this. The deer live without interference by man except that their territories are restricted; they are fed during the winter months and given salt so that they are accustomed to humans, but they are not driven or stalled. The herd is selectively culled which provides venison but it is not otherwise interfered with, this being a rare example of how man can live in close proximity to a species of large mammal without having the urge to exterminate it. This harmony is, however, only the result of hundreds of years of Royal protection which maintained the red deer for the sport of kings.

PÈRE DAVID'S DEER, *Elaphurus davidianus*

It may be more than two thousand years since the ancestors of this deer were common wild animals; their history is quite unknown except that some sub-fossil remains of Père David's deer have now been identified from archaeological sites in the East. It is a curious looking, rather heavily-built deer with a long tail, antlers that appear to be the wrong way round, for the tines point backwards, and wide-splayed hooves that suggest it evolved in a snow-clad landscape. Nothing is known, however, of the deer's natural environment for, although it is not a domesticated animal, until its discovery by the western world this deer survived only in the Imperial Hunting Park of Nan-Hai-Tze, south of Peking.

The deer, which is known in Chinese as the Mi-lou, was first seen by a European in 1865 when a Frenchman, Armand David, managed to gain entrance to the park which was surrounded by a wall 72 km in length. David succeeded in bringing the hides and bones of a pair of deer back to Paris and these immediately aroused the interest of zoos in Europe as well as the collecting fervour of the Duke of Bedford who was anxious to obtain these deer for his private collection of exotic animals at Woburn Abbey in Bedfordshire, England. During the next thirty years a number of deer were exported from Peking and they began to breed at Woburn. Meanwhile disaster struck the Nan-Hai-Tze deer for in 1894 a flooded river broke down the wall of the park and most of the herd escaped and were killed by the Chinese people who were at that time in a state of famine. Finally, during the Boxer rebellion, the few remaining deer were either killed or exported and by 1900 there were no Mi-lou left in China.

The survival of this species of deer was in great part due to the eleventh Duke of Bedford who managed to establish a breeding herd at Woburn Abbey. Despite set-backs in its numbers due to the two world wars, this herd has continued to flourish and the deer have now been exported to zoos all over the world, including to China. The story of Père David's deer is a classic example of how a species can be saved from extinction by a policy of management and breeding in captivity.

THE BLACKBUCK
FAMILY BOVIDAE, TRIBE ANTILOPINI

Antilope cervicapra. The Indian blackbuck belongs to the same group of bovids as the gazelles and over the course of history it has been of equal importance as a game animal. Also, like the red deer in Europe during the Middle Ages, blackbuck provided sport for the nobility. Not only were they hunted, but in addition and unlike red deer, the males were set against each other in fights. For this purpose the males were kept captive and cared for with the utmost attention to their health and feeding. Akbar, the Moghul emperor, not only hunted blackbuck with cheetahs on a grand scale but he owned at least one hundred animals for fighting. Each of these had its own rank and there was a complicated system for rating and matching of the bucks (Fig. 18.8, p. 167). There were special hospitals for the care of wounded animals and a stud for the training of tamed blackbuck that could be used as decoys to catch wild individuals as well as training them for fighting. Some blackbuck females became so tame that they could be milked but there is little evidence that they were bred in captivity in large numbers.

The blackbuck was until the last century the most common ungulate of the plains of India. It has been estimated that there were once at least four million on the subcontinent and herds of between 400 and 500 animals were not uncommon. As with all other game animals, however, the destruction of their habitat and the guns of the Europeans have combined to cause their demise and now they are only found in protected areas.

Remarkable as it may seem the countries where the blackbuck are now most abundant are the Argentine and Texas. In 1906 blackbuck, along with axis deer, fallow deer, red deer, and wild boar, were imported to a private estate in the southern pampas of the Argentine. The blackbuck thrived, mixing with the cattle, and have become so numerous that they have to be culled by the hundred. Similarly blackbuck were imported to Texas in the 1930s and 1940s and there are estimated to be at least 7000 animals in the Hill Country (Mungall, 1978). These Texan black-buck are held on ranches where they breed freely and mingle with other indigenous and exotic wildlife. They cannot be said to be domesticated but the herds are managed and selectively culled. They are hunted for sport and the meat, hides, and horns are valued commercially.

THE AFRICAN ANTELOPES
FAMILY BOVIDAE

In many countries of Africa there is at last an awareness of the great value of the large herds and diverse species of ungulates; not only are they of importance for tourism but their meat and hides can be exploited, and they are also necessary for the stability of the environment. The ranching of antelopes is now well established in East and South Africa and there are, for example, some 3000 ranches in the Transvaal where wild Springbok, *Antidorcas marsupialis*, are run with domestic cattle (Kyle, 1972). This is one way of exploiting the indigenous ungulates in a controlled manner; another is to go a stage further and to farm a few species of ruminants in a more restricted area, as is done at the Galana River estate in Kenya. Here oryx, *Oryx beisa*, buffalo, *Syncerus caffer*, and eland, *Taurotragus oryx*, are managed along with domestic cattle. They are dipped, castrated, de-horned, weighed, and in every way afforded the same treatment as the cattle (Fig. 18.9).

Up to the present time, however, the farming in this manner of indigenous bovids has not attracted large commercial interests and is only carried out by private enterprise. One difficulty is that these newly domesticated animals cannot be driven to a market or

Figure 18.9   Domesticated eland, *Taurotragus oryx*, Galana river estate, Kenya (photo P. A. Jewell).

slaughterhouse and so have to be killed at the farm. Problems such as this should be overcome, however, as it is in the best interests of both man and the environment to extend the range of these animals outside that of wildlife sanctuaries. Antelopes, equids, elephant, and all the great variety of African mammals have become adapted to life in the different regions of Africa over an immense period of time. This is in contrast to domestic cattle, originating in Europe, that are by nature grazers and browsers of temperate woodlands. Moreover domestic livestock have only been inhabitants of sub-Saharan Africa in any numbers since the fifteenth century AD and there are vast areas where their production rate is very low.

As producers of meat game animals have many advantages over cattle in the arid regions; they can range widely and therefore make the most of the low quality grazing whilst several species, particularly the oryx and the eland, require very little water. They are also resistant to the endemic diseases of Africa and will do well in the tsetse-infested areas where cattle will barely survive.

Cultural prejudices play a large part in the resistance of both Europeans and Africans to the ranching of game for meat. Every farmer believes he should own cattle, sheep, and goats, but if the increasing populations of the world are to go on eating meat these farmers must turn to the alternative sources.

THE ELAND, *Taurotragus oryx*

The eland is probably the most suitable of the African antelopes for experimental domestication, for this species has a social system that is based on a hierarchy of dominance in the male and not on territoriality. And indeed the eland is proving to be a highly successful domestic animal but principally, at present, in Russia, rather than in Africa. There are small herds of domestic eland that have been bred in Rhodesia (Posselt, 1963), and in Kenya (Kyle, 1972) but in the Askaniya-Nova Zoological Park in the Ukraine there is a healthy and established domestic stock. In 1968 according to Treus & Kravchenko there was a total of 408 eland that were all descended from four bulls and four cows that were obtained from Africa in 1892. Of these, 21 cows had been trained for milking in 1968 and the animals were husbanded in exactly the same manner as domestic cattle. The eland have acclimatized well to the conditions of the southern Ukraine; they are put out to graze in the spring and are stalled in the winter when they are fed on cereals, hay, and root crops as are cattle. The stock of domestic eland is now undergoing selective breeding for improvements in the quality of meat and the quantity of milk. Eland milk differs from that of cattle in having a higher proportion of fat and it also keeps fresh for a much longer period.

THE SAIGA ANTELOPE
FAMILY BOVIDAE, TRIBE SAIGINI

Saiga tatarica. The saiga has been hunted for its meat, hide, and

Figure 18.10   Saiga, *Saiga tatarica*

horns from the prehistoric period until the present day (Fig. 18.10). It is included here not because it has been domesticated but because it has been saved from extinction by Russian conservationists and the antelope are now so abundant in Russia that 6000 tonnes of meat are obtained from them each year (Linnard, 1963).

The saiga has always been hunted by the nomadic tribes of the Russian steppe lands both for its meat and for its horns which were believed to act as an aphrodisiac. According to Linnard Russia exported hundreds of thousands of horns every year to China for this purpose in the early nineteenth century. By the beginning of the present century, however, the formerly widespread saiga had become a rare animal. In 1919 a law was passed absolutely forbidding the hunting of saiga and the populations slowly increased again until in 1958 there were an estimated 540 000 saiga west of the Volga, and it is now the most abundant wild ungulate in the U.S.S.R.

Hunting of the saiga is still forbidden but there are official teams of hunters who selectively cull the animals on a commercial basis. In some areas 1000 saiga may be killed in a day but the hunting season is very short and care is taken that the stocks do not become permanently reduced.

Although its common name is the saiga antelope this bovid is not a true antelope but is closer in its affinities to the sheep and goats and it is classified in the subfamily Caprinae. It lives in the steppe lands and semi-desert of southern Russia.

THE MUSK OX
FAMILY BOVIDAE, TRIBE OVIBOVINI

Ovibos moschatus. The final species to be described in this chapter, the musk ox, has like other Northern ruminants, the saiga, the elk, and the reindeer, been hunted by man during and since the Pleistocene period (Fig. 18.11). It is an arctic mammal that, like the saiga, belongs to the subfamily Caprinae. It looks, however, a lot more like an ox than a sheep. The musk ox has the most northern distribution of all the ungulates and as such its use to man has obvious restrictions. Exploitation of the musk ox during the past and as an experimental domestic animal at the present day has been reviewed by Wilkinson (1972, 1975).

In the wild the musk ox is found only in the tundra zone of Canada and Greenland but experimental farms for the breeding of these bovids have been established in Alaska, Quebec, Norway and most recently in Siberia. The chief resource of the musk ox is its very fine and abundant underwool which can be plucked from the body when it is shed in the spring. The animals have proved easy to tame and as they are sedentary by nature they can be held in a restricted area. They are naturally gregarious and there is every possibility that they could make docile domesticated animals.

Figure 18.11   Musk ox,
Ovibos moschatus

Conclusions

The geography of domestication

Man was first set apart from the rest of the animal world by the intensive use of tools that were manufactured by his own hand. It may be argued that the next most important human accomplishment was the mastery of fire which opened up the way to an entirely new life. With the aid of fire humans could move north, survive in an Ice Age landscape, add to the variety of foods that they could eat, and preserve meat by smoking and drying it. Furthermore by setting light to large areas of forest whole herds of ungulates could be driven to their deaths.

Although man learned how to use tools and control fire tens of thousands of years before the invention of agriculture, these accomplishments should be seen as the first stages in the change from food procurement to food production. The next advance appears to have occurred first, towards the end of the Pleistocene period, in the Nile Valley and in the Near East where small groups of hunter-gatherers began to harvest and perhaps even to nurture wild cereals, including wheat and barley. For some thousands of years before this, however, it is likely that the sporadic cultivation of favoured plants took place, as well as the taming of individual animals. The end of the Pleistocene was the period of incipient agriculture and it was also the time of transition between tamed wolf and domestic dog.

To try to find relics of the first planted seeds or tamed animals is like trying to find the first stone tool. It is an impossible task and even if it were not it would bear little relevance to the understanding of the process of domestication. This can only be achieved by unravelling the ecological and cultural history of man combined with a study of the comparative patterns of human and animal behaviour. A species of mammal can only become domestic if it is able to adapt to living and breeding isolated within the human social system; this means that by nature it must conform broadly to the same kind of society. The animals must be gregarious rather than solitary and dominance must be established by means of a hierarchical or rank order rather than by the holding of exclusive territories.

Despite these strictures on behaviour what is surprising is that so few species of animal have been domesticated. In the New

World, until the arrival of the Spanish in the late fifteenth century AD, the North American Indians had only domesticated the dog and the turkey, whilst in South America by this time there were only three groups, dogs, camelids, and guinea pigs. No Australian animal has been domesticated (the dingo cannot be an endemic species) and the only contribution from the vast wealth of species in Africa to the list of domestic animals is the ass, possibly the cat, and the Guinea fowl. Why did the Australian aborigines not domesticate the kangaroo, nor the African Bushmen the eland, nor the North American Indians the bison or the mountain sheep?

Civilization has been built upon settlement, agriculture, and the keeping of livestock for food. It is therefore a ready assumption that the beginnings of this way of life were a natural and forward progression from a less satisfactory system. The present evidence suggests, however, that agriculture, and later domestication of livestock, were only adopted in a few restricted areas where the carrying capacity of the land, for one reason or another, could no longer support a stable population of hunter-gatherers. These reasons could be, and probably were, manifold and interlocking, and the impetus for change had very slow results.

The earliest centres for agriculture in the Old World, as known at present, are the Nile valley and the Near East, and New Guinea, whilst in the New World evidence for the earliest cultivation of plants has been found at somewhat later periods in Mexico and Peru. Although the change to a settled life was very slow, once achieved, it was irreversible and there is no archaeological evidence from anywhere in the world of a human society that has reverted to hunting as its sole means of subsistence. This is because farming is inevitably associated with demographic changes in the population; the mortality of the very young and the very old decreases when people cease to be nomadic because greater care can be given to them, family units become larger, and they become owners of houses and property with an established and restricted home range.

One aspect' in the process towards the husbanding of livestock that has been little emphasized in recent writings is that of the 'crop robbers' as described by Zeuner (1963). Wild grazing herbivores cannot be tolerated in the vicinity of carefully-nurtured plants. It is therefore very difficult to remain a hunter of large animals and cultivate crops at the same time, for any deer, gazelle, or bison (depending on the part of the world) that may wander near the farm must be driven ruthlessly away. This will naturally increase the flight distance of the wild animals and they will become more difficult to hunt. Only those species of ungulate that have the adaptive capacity to become controllable and closely associated with the human community can be tolerated. In this way domestication can be looked on as a process of adaptation in which the animals, say sheep and goats, moved into the new ecological niche provided by the early agriculturalists.

It is possible that the domestication of livestock occurred in this way in western Asia and it may also have happened like this with the camelids in South America but in North America agriculture became highly developed and widespread without the domestication of any mammal other than the dog. Yet wild sheep were as abundant in the mountainous regions of North America as in the Old World. This question has been discussed by Carr (1977) who has shown that although North American wild sheep have rather different ecological and behavioural habits from those of western Asia they were certainly hunted by man and there appears to be no straightforward reason why they did not become more intensively exploited.

The incentive to keep animals in a controlled manner as a 'walking larder', in preference to just hunting them in the wild when meat is required, clearly had no single origin. It resulted from a linkage of changing situations in which ecology, climate, natural resources, technology, and cultural factors all influenced each other. Once the incentive became established, however, it spread rather rapidly throughout the Old World with the exception of sub-Saharan Africa.

Evidence for the keeping of domestic animals in Africa during the prehistoric period has been reviewed by Shaw (1977). The spread of cattle, sheep and goats throughout the continent was very slow but it did occur and by the time that Europeans arrived in southern Africa in the fifteenth century AD, there were pastoral Hottentots who had large herds of long-horned cattle and fat-tailed sheep. The reasons for the slow spread, and in many regions total lack of agriculture and domestication in sub-Saharan Africa, can be attributed to the endemic diseases that attack humans and animals on this continent. Whereas in all other parts of the world the introduction of food production as opposed to food procurement has been associated with a rapid rise in the human population this is not so in Africa, because, here, many of the environmental changes that accompany agriculture also encourage disease. Clearance of land, irrigation, and the provision of standing water for animals increases the vectors for malaria, bilharzia (schistosomiasis), and river blindness (onchocerciasis). In addition sleeping sickness (trypanosomiasis) attacks both domestic cattle and humans and according to Shaw there are 10·5 million square kilometres of land in Africa that remain without cattle for this reason.

Human interference has altered the natural world far beyond the point of no return. Throughout the millions of years of the earth's history species of animals and plants have replaced each other in the natural course of evolution. The extermination by man of a very large number of organisms, large and small, may therefore be seen in evolutionary terms as a normal part of the success and failure of predator-prey relationships. On the other hand the manner in which man has moved the world's fauna and flora around is a unique and irreversible happening, the ecological and climatic consequences of which we may be only just on the edge of beginning to comprehend.

The earliest animal species to be moved by human agency was the wolf which in its domesticated form has accompanied man in his conquest of the whole world and has even travelled into outer space. As a wild species the wolf has lost the evolutionary fight as a competitive predator, but as the domesticated dog its biological success must be counted as high as that of the human species.

Other mammals that have, by their symbiotic relationship with man, achieved an almost equal biological success are the house mouse, the black and brown rats, the rabbit, and the cat as well as the common livestock animals – cattle, pigs, sheep and goats. Furthermore the distribution of the rabbit and the horse has not only been vastly increased but these mammals may indeed have been saved from extinction by the intervention of man. All these mammals are now so well established that, even if man were to disappear from the face of the earth, it is likely that they would continue to proliferate.

It is not only the domestic mammals that have been moved about by human agency. It appears that the rich and curious of all affluent societies have enjoyed keeping exotic animals and the Romans, particularly, made great efforts to reorganize the animal world to suit their own inclinations. Pliny*, for example, describes how an effort was made to introduce croaking frogs to those places that had none:

In Cyrene, the frogs were formerly dumb, and this species still exists, although croaking ones were carried over there from the continent. At the present day, even the frogs in the island of Seraphos are dumb; but when they are carried to other places, they croak; the same thing is also said to have taken place at Sicandrus, a lake of Thessaly.

(VIII, 83)

As is very well known, fashions change according to the mood and circumstances of society. Today there is a growing concern with the conservation of wild life, of the environment, and of its natural resources which is matched by their steady depletion. We look back on the era of big game hunting that prevailed throughout the last century with amazement and horror, and it is right to do so. However the greatest effect that man has had on the living world is perhaps not so much in the outright extinction of large animals such as the mammoth and the dodo but in the loss of diversity. After all, amazingly, there are still whales in the sea, elephants on the land, and turtles in the lakes, but there are also vast areas of the world where the numbers and varieties of both plants and animals have been so reduced that the landscape has been transformed.

Five thousand years ago western Asia was a land of fertile

* Translated by Bostock & Riley (1855, II, p. 352).

plains, it was the hearth of civilization; today these plains are mostly desert, due to centuries of over-grazing by goats and sheep and to the relentless collection of every scrap of shrub and twig for firewood and the production of charcoal. Climatic change may also have hastened the production of deserts in the Middle East, as in the Sahara and elsewhere, but then it is not known to what extent the activities of man have, in turn, affected climate and rainfall. Deforestation and the extensive burning of fires have altered the atmosphere and very likely resulted in small changes of climate which could escalate in the future, if, for example, the forests of South America and South East Asia continue to be destroyed at their present rate.

The loss of diversity within wild faunas and floras that has been steadily increasing since the first spread of agriculture is now becoming evident with domesticated species as well, and is as much to be deplored. Since animals were first domesticated they have become adapted to different climates and to different conditions of feeding and housing so that a multitude of breeds has been developed all over the world within each species. These breeds often differed markedly within very small areas, as for example with, say, the separate kinds of cattle found in north and south Devon in England. Nowadays, however, with commercial breeding and artificial insemination, farming is taking on an entirely new look and local breeds of livestock are becoming a rarity.

The result is that the classes of livestock are becoming much less varied and there are a few favoured breeds that are almost ubiquitous. Friesian cattle, for example, with their stark black and white markings, can be seen throughout the 'developed' world. The lack of diversity within domesticated animals has potential dangers, particularly for cattle because their wild progenitor is extinct and once genetic material is lost from the species it cannot be replaced.

It is becoming just as important to conserve the old, well-established breeds of domestic animals as it is to conserve wild species, and it is for this reason that societies such as the *Rare Breeds Survival Trust* of Britain are attempting to preserve the remaining unimproved breeds of British livestock (Alderson, 1978). Their loss, in the long term, could only reduce both the profit and pleasure of future generations of farmers.

In any attempt to trace the history of man's exploitation of the animal world we have to remember that we are dealing with the human being, a species once termed 'the irascible ape', who at the same time is not an ape because of his powers of conceptual thought and his capacity to plan for the future. Man has changed the world and will continue to do so. There is no easy solution to the problems of over-population and the needs of the hungry, but if man is irascible he is, as well, endlessly ingenious and adaptable. He is also an artist who, since the Palaeolithic period, has shown his appreciation of the beauty and diversity of form to be found in living animals, and we must no longer continue with the extermination of these animals. Despite the commercial greed that holds an unrelenting grip on modern societies, we must believe that altruism will prevail, that some wild places will remain, and that we will learn how to manage and protect an increasing number of wild animals and plants that at present appear doomed. This can be achieved, if, instead of encroaching on the last reserves of the natural environment in efforts to provide more farmland, the indigenous faunas and floras are exploited. Additional animals will have to be domesticated and the elephant, the zebra, and the eland must join the ox and the sheep as providers of meat, leather and other necessities for the welfare of the master predator.

Appendix I

Nomenclature of the domestic mammals

The tenth edition of the *Systema Naturae* written by Linnaeus and published in 1758 is internationally accepted as the basis for zoological nomenclature. In this work, written in Latin, Linnaeus gave names, descriptions, and localities to more than 4000 organisms, amongst which were included all the common domestic animals (Linnaeus, 1758). It is now realized, however, that there are a number of inconsistencies and difficulties that make the application of the Linnaean system of binomials to domestic animals rather awkward. This is because our ideas of what is a species have changed and become more complicated. We can no longer say that a dog is a different species to a wolf just because it looks so different, as Linnaeus could. Because of difficulties such as this it has been suggested that domestic mammals should be excluded from formal zoological nomenclature, but up to the present no agreement has been reached on what system should take its place. Therefore an attempt will be made here to give a short summary of the problems and to list the names of the domestic mammals and their wild parent species, as they have been used in this book.

A domesticated Indian elephant can be given the same name (*Elephas maximus* L.) as its wild counterpart because they will both belong to the same breeding population. The problems arise when formal nomenclature is applied to the domestic mammals whose reproduction is so closely controlled by man that the populations within the domestic and wild groups are completely separated. There are 21 domestic mammals that can be said to fall into this category and 19 of them were named by Linnaeus, as shown in the Table. Within this group of 19 taxa there are three nomenclatorial problems that have taxonomic implications. When Linnaeus was familiar with both the wild and the domestic form of a species and they looked alike, as with his native reindeer, he gave them the same name, *Cervus tarandus* now called *Rangifer tarandus*. On the other hand because he failed to see the relationship between the wolf and the dog he gave them the separate species names, *Canis lupus* and *Canis familiaris*. With yet others, for example the goats and sheep, he had no knowledge of the wild ancestor and so he named only the domestic form. Implicit in these three examples is the still unresolved problem of whether or not domestic animals should be treated as taxonomically identical with their wild progenitors.

In order to overcome these difficulties several different systems of nomenclature have been devised for the domestic mammals but none has so far received international recognition. The most widely accepted system is that proposed by Bohlken (1961), it being much favoured by archaeozoologists particularly in Germany. Bohlken's solution is to call the domestic form by the first available name for the wild species, followed by the linking word 'forma' (f.) and then by the earliest name, according to the rule of priority, for the domestic animal. In this way we would have *Canis lupus* f. *familiaris* L. for the dog and *Capra aegagrus* f. *hircus* L. for the domestic goat. This arrangement is, however, clumsy and it has the disadvantage that it assumes certain identification of the wild progenitor which for some domestic animals, for example the ferret, may never be established.

The domestic mammals, although they are fully fertile with their wild progenitors, do not normally breed with them. Therefore they may be considered as separate, reproductively isolated populations, although they are not distinct species in the true taxonomic sense. For purposes of clarity these domestic mammals require names that are consistent and which by tradition and common usage are known to everyone. The obvious names to use are those that hold priority by being the oldest, and this is the practice that has been adopted here, with where necessary the next available name being appropriated for the wild species (see the table). Thus the domestic water buffalo is called by the name by which it was first described (by Linnaeus), this being *Bubalus bubalis* (L.).* In the past this name has also been used for the wild water buffalo, but as Linnaeus knew nothing of this animal it is more appropriate to give it the next available name based on a description of the wild species, which is *Bubalus arnee* (Kerr, 1792). This arrangement follows the now widely accepted premise that names based on descriptions of domestic mammals should not be used for wild species whilst at the same time keeping as close as possible to the traditional nomenclature.

As can be seen from the Table there are three mammals in which the same name is retained for the domestic form and the wild species, the rabbit, the reindeer, and the Bali cattle. This is because there is no tradition of naming them separately and to create new binomials for the domestic forms at this juncture would only add to the nomenclatorial confusion. If at some future time there is international agreement on a system of formal nomenclature for the domestic animals as a whole, and this would have to include the birds and fish, then it would probably be necessary to create a small number of new names to cover these anomalies.

* L = Linnaeus, 1758. Brackets around an author's name, following the Latin binomial for a species, indicates that the genus has been changed. In this case Linnaeus named the water buffalo *Bos bubalis* but the genus has since been changed to *Bubalus*, a name that was first used by Hamilton Smith in 1827.

Latin names of domestic mammals and their wild parent species as used in this book

NOTE   The distributions are broad assessments of what was probable for the wild species at the time of their first domestication.
* ABBREVIATION   L = Linnaeus, 1758.
† For explanation of the brackets see the footnote, p. 195.

Wild progenitor	Domestic form
ORDER LAGOMORPHA	
Oryctolagus cuniculus (L*)†, Rabbit Probably N W Africa and Spain	*Oryctolagus cuniculus* (L), Rabbit (no separate name)
ORDER RODENTIA	
Cavia – species not known, Cavy South America	*Cavia porcellus* (L), Guinea pig
ORDER CARNIVORA	
Canis lupus L, Wolf Widespread through the northern hemisphere except Africa	*Canis familiaris* L, Dog
Mustela putorius L, European polecat Europe	
Mustela eversmanni Lesson, 1827, Steppe polecat Eastern Europe and Asia	*Mustela furo* L, Ferret
Felis silvestris Schreber, 1777, Cat Europe, Asia, Africa	*Felis catus* L, Cat
ORDER PERISSODACTYLA (Odd-toed Ungulates)	
Equus ferus Boddaert, 1785, Horse Southern Russia and Central Asia	*Equus caballus* L, Horse
Equus africanus (Fitzinger, 1857), Ass North Africa and possibly western Asia	*Equus asinus* L, Donkey
ORDER ARTIODACTYLA (Even-toed Ungulates)	
Sus scrofa L, Boar Europe, Asia, and North Africa	*Sus domesticus* Erxleben, 1777, Pig

Wild progenitor	Domestic form
Possibly *Lama guanicoe* (Muller, 1776), Guanaco South America, semi-deserts and high altitudes	*Lama glama* (L), Llama
Lama – species not known South America	*Lama pacos* (L), Alpaca
Camelus – species not known Asia or possibly North Africa	*Camelus dromedarius* L, Dromedary or one-humped camel
Camelus ferus Przewalski, 1883, Bactrian camel Central Russia and Asia, dry steppe and desert	*Camelus bactrianus* L, Bactrian or two-humped camel
Rangifer tarandus (L), Reindeer Northern Europe and Asia, and North America	*Rangifer tarandus* (L), Reindeer (no separate name)
Bubalus arnee (Kerr, 1792), Water buffalo India, southern Asia, and possibly western Asia, wet areas	*Bubalus bubalis* (L), River and Swamp buffaloes
Bos primigenius Bojanus, 1827, Aurochs or Giant ox (extinct) Widespread throughout Europe, Asia, and North Africa	*Bos taurus* L, European cattle *Bos indicus* L, Indian humped cattle or Zebu
Bos mutus (Przewalski, 1883), Yak Mountains of Tibet, Nepal, and the Himalayas	*Bos grunniens* L, Yak
Bos gaurus H. Smith, 1827, Gaur India and S E Asia	*Bos frontalis* Lambert, 1804, Mithan or Gayal
Bos javanicus d'Alton, 1823, Banteng Borneo and the Islands of S E Asia	*Bos javanicus* d'Alton, 1823, Bali cattle (no separate name)
Capra aegagrus Erxleben, 1777, Bezoar goat Mountains of western Asia	*Capra hircus* L, Goat
Ovis orientalis Gmelin, 1774, Asiatic mouflon Mountains of western Asia	*Ovis aries* L, Sheep

Appendix II

Climatic sequences and archaeological divisions of the Quaternary period

Period	Glacial and Interglacial* Stages			
	N. AMERICA	N. EUROPE	ALPS	BRITAIN
Quaternary Post Pleistocene (Holocene)	Post Glacial			Flandrian
Upper Pleistocene	Late Glacial			
	Wisconsian	Weichsel	Würm	Devensian
	*Sangamon**	*Eemian*	*Riss-Würm*	*Ipswichian*
	Illinoian	Saale	Riss	Wolstonian
Middle Pleistocene	*Yarmouth*	*Holstein*	*Mindel-Riss*	*Hoxnian*
	Kansan	Elster	Mindel Complex	Anglian
	Aftonian	*Cromerian Complex*	*Günz-Mindel*	*Cromerian*

* Interglacial stages in *italics*

Pollen Zones	Climate	Time before present	Archaeological Divisions	
BRITAIN	N. EUROPE		N. EUROPE	WESTERN ASIA (LEVANT)

COLD ← | → WARM

VIII Sub-Atlantic			Neolithic-Modern	Bronze Age- Modern
VIIb Sub-Boreal		2500	Neolithic	
VIIa Atlantic		5000		Neolithic
VI Later Boreal		7500		
		9000		
V Early Boreal			Mesolithic	Pre-pottery Neolithic B
IV Pre-Boreal		9600		Pre-pottery Neolithic A
		10 300		
			Late Glacial	Proto-Neolithic (Jericho only)
III Younger Dryas		10 800		
II Allerød Interstadial		12 000	Upper Palaeolithic	Natufian
I Older Dryas		14 000		Kebaran (not Jericho)
		50 000		
		100 000?		
		200 000?	correlations increasingly uncertain	
		300 000?		
		400 000?		
		500 000?		

References and publications for further reading

AJAYI, S 1975. *Domestication of the African giant rat*. Department of Forest Resources, University of Ibadan, Ibadan, Nigeria.

ALDERSON, L 1978. *The chance to survive. Rare breeds in a changing world*. Cameron and Tayleur, David and Charles, England.

AMOROSO, E C AND JEWELL P A 1963. The exploitation of the milk-ejection reflex by primitive peoples. In MOURANT, A E AND ZEUNER, F E (eds.) Man and cattle. *Roy. anthrop. Inst. Occ. Paper* 18, 126–138.

BACHRACH, M 1947. *Fur*. Prentice Hall, New York.

BAHN, P G 1978. The 'unacceptable face' of the West European Upper Palaeolithic. *Antiquity* 52, 183–192.

BASKIN, L M 1974. Management of the ungulate herds in relation to domestication. In GEIST, V AND WALTHER, F (eds.) *The behaviour of ungulates and its relation to management. I U C N Publications*, N.S. 24 2, 530–542.

BATE, D M A 1937. Part II Palaeontology: the fossil fauna of the Wady el-Mughara caves. In GARROD, D A E AND BATE, D M A *The stone age of Mount Carmel I*. Clarendon Press, Oxford, 139–240.

BERTRAM, C K R AND BERTRAM, G C L 1968. The Sirenia as aquatic meat-producing herbivores. *Symp. zool. Soc. Lond.* 21, 385–391.

BERTRAM, G C L AND BERTRAM, C K R 1963. The status of manatees in the Guianas. *Oryx* 7, 90–93.

BEWICK, T 1790. *A general history of quadrupeds*. E. Walker, Newcastle upon Tyne.

BLAXTER, K L, KAY, R N B, SCHARMAN, G A M, CUNNINGHAM, J M M AND HAMILTON, W J 1974. *Farming the Red Deer*. Department of Agriculture and Fisheries for Scotland, Edinburgh.

BOHLKEN, H 1961. Haustiere und zoologische Systematik. *Z. Tierzücht. u. Züchtungsbiol.* 76, 107–113.

BÖKÖNYI, S 1968. *Data on Iron Age horses of Central and Eastern Europe*. Am. Sch. prehist. Res., Peabody Museum, Harvard University, Bull. 25.

BÖKÖNYI, S 1974. *History of domestic mammals in Central and Eastern Europe*. Akadémiai Kiadó, Budapest.

BÖKÖNYI, S 1975. Vlasac: an early site of dog domestication. In CLASON, A T (ed.) *Archaeozoological Studies*. North Holland, Amsterdam, 167–178.

BÖKÖNYI, S 1976. Development of early stock rearing in the Near East. *Nature*, 264, 5581, 19–23.

BOSTOCK, J AND RILEY, H T (transl.) 1855. *The natural history of Pliny* 5 vols. Bohn's Classical Library, Henry G. Bohn, London.

BURLEIGH, R, CLUTTON-BROCK, J, FELDER, P J AND SIEVEKING, G DE G 1977. A further consideration of Neolithic dogs with special reference to a skeleton from Grime's Graves (Norfolk), England. *J. arch. Sci.* 4, 4, 353–365.

CAMPBELL, B G 1972 Man for all seasons. In CAMPBELL, B G (ed) *Sexual selection and the descent of man 1871–1971*. Heinemann, London.

CAMPBELL, B G 1976. *Humankind emerging*. Little, Brown and Co., Canada and USA.

CARR, C 1977. Why didn't the American Indians domesticate sheep? In REED, C A (ed.) *Origins of agriculture*. Mouton, Hague, Paris, 637–693.

CLASON, A T 1978. Late Bronze Age–Iron Age zebu cattle in Jordan? *J. arch. Sci.* 5, 1, 91–94.

CLUTTON-BROCK, J 1974. The Buhen horse. *J. arch. Sci.* 1, 1, 89–100.

CLUTTON-BROCK, J 1979. The mammalian remains from the Jericho Tell. *Proc. prehist. Soc.* 45, 135–158.

CLUTTON-BROCK, J AND UERPMANN, H-P 1974. The sheep of early Jericho. *J. arch. Sci.* 1, 261–274.

COCKRILL, W R 1967. The water buffalo. *Scient. Amer.* 217, 118–125.

COCKRILL, W R (ed.) 1974. *The husbandry and health of the domestic buffalo.* Rome, FAO.

COHEN, H N 1977. *The food crisis in prehistory.* Yale University Press, New Haven.

COMPAGNONI, B AND TOSI, M 1978. The camel: its distribution and state of domestication in the Middle East during the third millennium BC in the light of finds from Shahr-I Sokhta. In MEADOW, R H AND ZEDER, M A (eds.) *Approaches to faunal analysis in the Middle East.* Peabody Museum Bulletin 2, 91–103.

CORBET, G B 1978. *The mammals of the Palaearctic region – a taxonomic review.* Trustees of the British Museum (Natural History), London.

CRAWFORD, M A 1968. Comparative nutrition of wild animals. *Symp. zool. Soc. Lond.* 21.

DARWIN, C 1859. *On the origin of species by means of natural selection or the preservation of favoured races in the struggle for life.* Murray, London.

DAVIES, N M AND GARDINER, A H 1936. *Ancient Egyptian Paintings.* University of Chicago Press.

DAVIS, S J M AND VALLA, F R 1978. Evidence for domestication of the dog 12 000 years ago in the Natufian of Israel. *Nature* 276, No. 5688, 608–610.

DIGBY, B 1926. *The mammoth and mammoth-hunting in north-east Siberia.* Witherby, London.

EPSTEIN, H 1969. *Domestic animals of China.* Commonwealth Agricultural Bureaux, England.

EPSTEIN, H 1971. *The origin of the domestic animals of Africa.* Africana Publishing Corporation, New York. 2 vols.

EPSTEIN, H 1977. *Domestic animals of Nepal.* Holmes & Meier, New York.

EWART, J C 1907. On skulls of horses from the Roman fort at Newstead, near Melrose, with observations on the origin of domestic horses. *Trans. R. Soc. Edinb.* 45, 555–588.

FLOWER, W H AND LYDEKKER, R 1891. *An introduction to the study of mammals, living and extinct.* Adam and Charles Black, London.

FORSTER, E S AND HEFFNER, E H (transl.) 1968. *Lucius Junius Moderatus Columella on agriculture.* Vol. II. Loeb Classical Library No. 407. Heinemann, London.

FRÄDRICH, H 1974. A comparison of behaviour in the Suidae. In GEIST, V AND WALTHER, F (eds.) *The behaviour of ungulates and its relation to management. I U C N Publications* N.S. 24, 1, 133–144.

FRANKLIN, W L 1974. The social behavior of the vicuna. In GEIST, V AND WALTHER, F (eds.) *The behaviour of ungulates and its relation to management. I U C N Publications* N.S. 24, 1, 477–488.

GALTON, F 1865. The first steps towards the domestication of animals. *Trans. ethnolog. Soc. Lond.* N.S. 3, 122–138. Reprinted in *Inquiries into human faculty.* J. M. Dent, London 1907.

GAUTHIER-PILTERS, H 1974. The behaviour and ecology of camels in the Sahara, with special reference to nomadism and water management. In GEIST, V AND WALTHER, F (eds.) *The behaviour of ungulates and its relation to management. I U C N Publications* N.S. 24, 2, 542–552.

GEIST, V 1971. *Mountain sheep, a study in behaviour and evolution.* University of Chicago Press, Chicago.

GEIST, V 1975. *Mountain sheep and man in the northern wilds.* Cornell University Press, New York.

GEIST, V AND WALTHER, F (eds.) 1974. *The behaviour of ungulates and its relation to management*. *I U C N Publications* N.S. **24**, 2 vols.

GRAY, A P 1972 *Mammalian hybrids*. Commonwealth Agricultural Bureaux, England.

HARCOURT, R 1974. The dog in prehistoric and early historic Britain. *J. archaeol. Sci.* **1**, 2, 151–175.

HARRIS, W C 1839. *The wild sports of southern Africa*. John Murray, London.

HENRY, D 1975. Fauna in the Near Eastern archaeological deposits. In WENDORF, F AND MARKS, A E (eds.) *Problems in prehistory: North Africa and the Levant*. S M U Press, Dallas, 379–385.

HERRE, W 1952. Studien über die wilden und domestizierten Tylopoden Sudamerikas. *Zool. Gart.* **19**, 70–98.

HERRE, W 1962. Ist *Sus (Porcula) salvanius* Hodgson, 1847 eine Stammart von Hausschweinen? *Z. Tierzüct. und Züchtungsbiol.* **76**, 265–281.

HERRE, W AND RÖHRS, M 1977. Zoological considerations on the origins of farming and domestication. In REED, C A (ed.) *Origins of agriculture*. Mouton Publishers, Hague, Paris, 245–280.

HIGGS, E S (ed.) 1972. *Papers in economic prehistory*. Cambridge University Press, Cambridge.

HIGGS, E S (ed.) 1975. *Palaeoeconomy*. Cambridge University Press, Cambridge.

HILZHEIMER, M 1941. *Animal remains from Tell Asmar*. Studies in Ancient Oriental Civilization no. 20, University of Chicago Press.

HOOPER, W D AND ASH, H B (transl.) 1967. *Marcus Porcius Cato on agriculture. Marcus Terentius Varro on agriculture*. Loeb Classical Library No. 283, Heinemann, London.

HOPF, M 1969. Plant remains and early farming in Jericho. In UCKO, P J AND DIMBLEBY, G W *The domestication and exploitation of plants and animals*. Duckworth, London, 361–381.

INGERSOLL, D, YELLEN, J E AND MACDONALD, W (eds.) 1977. *Experimental archeology*. Columbia University Press, New York.

JAMIESON, A 1966. The distribution of transferrin genes in cattle. *Heredity* **21**, 191–328.

JARMAN, M R 1972. European deer economies and the advent of the Neolithic. In HIGGS, E S (ed.) *Papers in economic prehistory*. Cambridge University Press, Cambridge, 125–149.

JARMAN, M R AND WILKINSON, P F 1972. Criteria of animal domestication. In HIGGS, E S (ed.) *Papers in economic prehistory*. Cambridge University Press, Cambridge, 83–97.

JEWELL, P A 1966. The concept of home range in mammals. In JEWELL, P A AND LOIZOS, C *Play, exploration and territory in mammals*. *Symp. zool. Soc. Lond.* **18**, 85–107.

JEWELL, P A, MILNER, C AND MORTON-BOYD, J 1974. *Island survivors – the ecology of the Soay sheep of St. Kilda*. University of London, Athlone Press, London.

JONES, H L (transl.) 1969. *The geography of Strabo*. Vol. 2. Loeb Classical Library No. 50, Heinemann, London.

KENYON, K M 1957. *Digging up Jericho*. Ernest Benn, London.

KING, J E 1953. Mammal bones from Khirokitia and Erimi. In DIKAIOS, P *Khirokitia*. App. III. Oxford University Press, Oxford, 431–437.

KIPLING, R 1902. *Just so stories*. Macmillan, London.

KLEIN, R G 1973. *Ice-Age hunters of the Ukraine*. University of Chicago Press, Chicago.

KLEIN, R G 1974. Ice-Age hunters of the Ukraine. *Scient. Am.* **230**, 6, 96–105.

KLINGEL, H 1974. A comparison of the social behaviour in the Equidae. In GEIST, V AND WALTHER, F (eds.) *The behaviour of ungulates and its relation to management*. *I U C N Publications* N.S. **24**, 1, 124–133.

KOPPER, J S AND WALDREN, W 1967. Balearic prehistory: a new perspective. *Archaeology* **20**, 2, 108–115.

KOWALSKI, K 1967. The Pleistocene extinction of mammals in Europe. In MARTIN, P S AND WRIGHT, H E (eds.) *Pleistocene extinctions*. Yale University Press, New Haven, 349–365.

KRUUK, H 1972. Surplus killing by carnivores. *J. zool. Soc. Lond.* **166**, 233–244.

KURTÉN, B 1968. *Pleistocene mammals of Europe*. The World Naturalist, Weidenfeld and Nicolson, London.

KYLE, R 1972. *Meat production in Africa – the case for new domestic species*. University of Bristol, Bristol.

LANE FOX, R 1973. *Alexander the Great*. Allen Lane, London.

LAYARD, A H 1849. *Nineveh and its remains*. John Murray, London.

LEE, R B 1968. What hunters do for a living, or how to make out on scarce resources. In LEE, R B AND DE VORE, I (eds.) *Man the hunter*. Aldine, Chicago, 30–48.

LEE, R B 1969. 'Kung bushmen subsistence, an input-output analysis. In VAYDA, A P (ed.) *Environment and cultural behaviour*. Natural History Press, New York, 47–79.

LEE, R B 1979. *The 'kung San: men, women and work in a foraging society*. Cambridge University Press.

LEE, R B AND DE VORE, I (eds.) 1968. *Man the hunter*. Aldine, Chicago.

LEGGE, A J 1972. Prehistoric exploitation of the gazelle in Palestine. In HIGGS, E S (ed.) *Papers in economic prehistory*. Cambridge University Press, Cambridge, 119–125.

LEGGE, A J 1977. The origins of agriculture in the Near East. In MEGAW, J V S (ed.) *Hunters, gatherers, and first farmers beyond Europe*. Leicester University Press, Leicester, 51–69.

LINNAEUS, C 1758. *Systema Naturae*. Facsimile edition 1958. Trustees of the British Museum (Natural History), London.

LINNARD, W 1963. The saiga. *Oryx* 7, 30–33.

LUMLEY, H DE 1969. Une cabane de chasseuse acheuléens dans la grotte du Lazaret à Nice. *Archeologia* 28, 26–33.

MARTIN, P S AND WRIGHT, H E 1967. *Pleistocene extinctions*. Yale University Press, New Haven.

MECH, L D 1970. *The wolf: the ecology and behaviour of an endangered species*. New York.

MEGAW, J V S (ed.) 1977. *Hunters, gatherers, and first farmers beyond Europe*. Leicester University Press, Leicester.

MELLAART, J 1975. *The Neolithic of the Near East*. Thames and Hudson, London.

MENDELSSOHN, H 1974. The development of the populations of gazelles in Israel and their behavioural adaptations. In GEIST, V AND WALTHER, F (eds.) *The behaviour of ungulates and its relation to management. I U C N Publications* N.S. **24**, 2, 722–744.

MOHR, E 1971. *The Asiatic wild horse*. J A Allen & Co. London.

MORRISON-SCOTT, T C S 1952. The mummified cats of Ancient Egypt. *Proc. zool. Soc. Lond.* **121**, 4, 861–867.

MOURANT, A E AND ZEUNER, F E 1963. Man and cattle. *Roy. anthrop. Inst. Occ. Paper* **18**.

MUNGALL, E C 1978. *The Indian blackbuck antelope: a Texas view*. Kleberg Studies in Natural Resources, Texas.

NADLER, C F, KOROBITSINA, K V, HOFFMAN, R S AND VORONTSOV, N N 1973. Cytogenic differentiation, geographic distribution and domestication in Palaearctic sheep (*Ovis*). *Z. Säugetierk* **38**, 2, 109–125.

NEWBERRY, P E 1893. *Beni Hasan*, part 1. In GRIFFETH, F L (ed.) *Archaeological survey of Egypt*. Egypt Exploration Fund, Kegan Paul, Trench, Trübner Co., London.

OWEN, R 1846. *A History of British Fossil Mammals and Birds*. John Van Voorst, London.

PFEFFER, P 1967. Le mouflon de Corse (*Ovis ammon musimon*); position systematique, écologie et éthologie comparées. *Mammalia* **31**, 1–262.

PIETTE, E 1906. Le chevêtre et la semi-domestication des animaux aux temps pléistocènes. *L'Anthropologie* **17**, 27–53.

PIRES-FERREIRA, J W, PIRES-FERREIRA, E AND KAULICKE, P 1976. Preceramic animal utilization in the Central Peruvian Andes. *Science* **194**, 483–490.

POCOCK, R I 1907. On English domestic cats. *Proc. zool. Soc. Lond.*, 143–168.

POCOCK, R I 1951. *Catalogue of the genus* Felis. Trustees of the British Museum (Natural History), London.

POLLOCK, N C 1969. Some observations on game ranching in Southern Africa. *Biol. Conserv.* **2**, 1, 18–24.

POSSELT, J 1963. The domestication of the eland. *Rhod. J. agric. Res.* **2**, 81–89.

RAWLINSON, G (transl.) 1964. *The histories of Herodotus.* 2 vols. Everyman's Library No. 405. Dent, London.

READ, C 1925. *The origin of man* (2nd ed.). Cambridge University Press, Cambridge.

REED, C A (ed.) 1977. *Origins of agriculture.* Mouton, Hague, Paris.

RENFREW, J M 1973. *Palaeoethnobotany – The prehistoric food plants of the Near East and Europe.* Methuen, London.

RUDENKO, S I 1970. *The frozen tombs of Siberia.* THOMPSON, M W (transl.). J M Dent & Sons, London.

SAHLINS, M 1972. *Stone age economics.* Aldine, Chicago.

SANSON, A 1869. Nouvelles déterminations des espèces chevalines du genre *Equus. c.r. hebd. Séanc. Acad. Sci. Paris* **69**, 1204–1207.

SCHALLER, G B 1977. *Mountain monarchs, wild sheep and goats of the Himalaya.* University of Chicago Press, Chicago.

SCHENKEL, R 1967. Submission: its features and function in the wolf and dog. *Am. Zool.* **7**, 319–329.

SCOTT, J P 1950. The social behaviour of dogs and wolves: an illustration of sociobiological systematics. *Ann. N.Y. Acad. Sci.* **51**, 1000–1021.

SCOTT, J P 1967. The evolution of social behaviour in dogs and wolves. *Am. zool.* **7**, 373–381.

SCULLARD, H H 1974. *The elephant in the Greek and Roman world.* Thames and Hudson, London.

SHAW, T 1977. Hunters, gatherers and first farmers in West Africa. In MEGAW, J V S (ed.) *Hunters, gatherers and first farmers beyond Europe.* Leicester University Press, Leicester, 69–127.

SHORT, R V 1976. The introduction of new species of animals for the purpose of domestication. *Symp. zool. Soc. Lond.* **40**, 321–333.

SIKES, S K 1971. *The natural history of the African elephant.* Weidenfeld & Nicolson, London.

SIMOONS, J 1968. *A ceremonial ox of India. The mithan in nature, culture, and history.* University of Wisconsin Press, Wisconsin.

SIMOONS, F J 1979. Dairying, milk use and lactose malabsorption in Eurasia: a problem in culture history. *Anthropos* **74**, 61–80.

STUDER, T 1901. Die prähistorischen Hunde in ihrer Beziehung zu den gegenwärtig lebenden Rassen. *Abh. schweiz. paläont. Ges.* **28**, 1–137.

STURDY, D A 1972. The exploitation patterns of a modern reindeer economy in west Greenland. In HIGGS, E S (ed.) *Papers in economic prehistory.* Cambridge University Press, Cambridge, 161–169.

STURDY, D A 1975. Some reindeer economies in prehistoric Europe. In HIGGS, E S (ed.) *Palaeoeconomy.* Cambridge University Press, Cambridge, 55–97.

TITCOMB, M 1969. *Dog and man in the ancient Pacific.* Bernice P. Bishop Museum Special Publication, **59**, Honolulu, Hawaii.

TODD, N B 1977. Cats and commerce. *Scient. Am.* **237**, 5, 100–107.

TREUS, V AND KRAVCHENKO, D 1968. Methods of rearing and economic utilization of eland in Ashaniyanova Zoological Park. *Symp. zool. Soc. Lond.* **21**, 395–411.

TURNBULL, P F AND REED, C A 1974. The fauna from the terminal Pleistocene of Palegawra Cave, a Zarzian occupation site in north-eastern Iraq. *Fieldiana Anthropology* **63**, 3, 81–146.

TURNER, P (ed.) 1962. *Pliny's Natural History in Philemon Holland's translation.* Centaur Press, London.

TWIGG, G I 1978. The role of rodents in plague dissemination: a worldwide review. *Mammal Review* 8, 3, 77–110.

VAN GELDER, R G 1969. *Biology of Mammals.* Charles Scribner's Sons, New York.

VAN GELDER, R G 1979. Comments on a request for a declaration modifying Article 1 so as to exclude names proposed for domestic animals from zoological nomenclature. (Z.N.(5)1935) *Bull. zool. Nom.* 36, 1, 5–9.

VERESHCHAGIN, N K 1974. The mammoth 'cemeteries' of north-east Siberia. *Polar Record* 17, 106, 3–12.

WATSON, J S (transl.) 1884. *Xenophon's minor works.* Bohn's Classical Library. George Bell & Sons, London.

WEIR, B J 1974. Notes on the origin of the domestic guinea pig. In ROWLANDS, I W AND WEIR, B J (eds.) The biology of hystricomorph rodents. *Symp. zool. Soc. Lond.* 34, 437–446.

WHEAT, J B 1967. A Palaeo-Indian bison kill. *Scient. Amer.* 216, 44–52.

WILKINSON, P F 1972. Current experimental domestication and its relevance to prehistory. In HIGGS, E S (ed.) *Papers in economic prehistory.* Cambridge University Press, Cambridge, 107–119.

WILKINSON, P F 1975. The relevance of musk ox exploitation to the study of prehistoric animal economies. In HIGGS, E S (ed.) *Palaeoeconomy.* Cambridge University Press, Cambridge, 9–53.

YAZAN, Y AND KNORRE, Y 1964. Domesticating elk in a Russian national park. *Oryx* 7, 301–304.

YELLEN, J E 1977. *Archaeological approaches to the present. Models for reconstructing the past.* Academic Press, New York.

ZARINS, J 1976. *The domestication of Equidae in third millennium BC Mesopotamia.* Thesis no. T-26263, Joseph Regenstein Library, University of Chicago, Illinois.

ZEUNER, F E 1953. The colour of the wild cattle of Lascaux. *Man* 53, 68–69.

ZEUNER, F E 1958. Dog and cat in the Neolithic of Jericho. *Pal. Expl. Quat. Lond.*, 52–55.

ZEUNER, F E 1963. *A history of domesticated animals.* Hutchinson, London.

ZHIGUNOV, P S (ed.) 1968. *Reindeer husbandry.* Translated from the Russian by the Israel Program for Scientific Translations, Jerusalem.

Index*

* Page numbers in italic indicate figures.

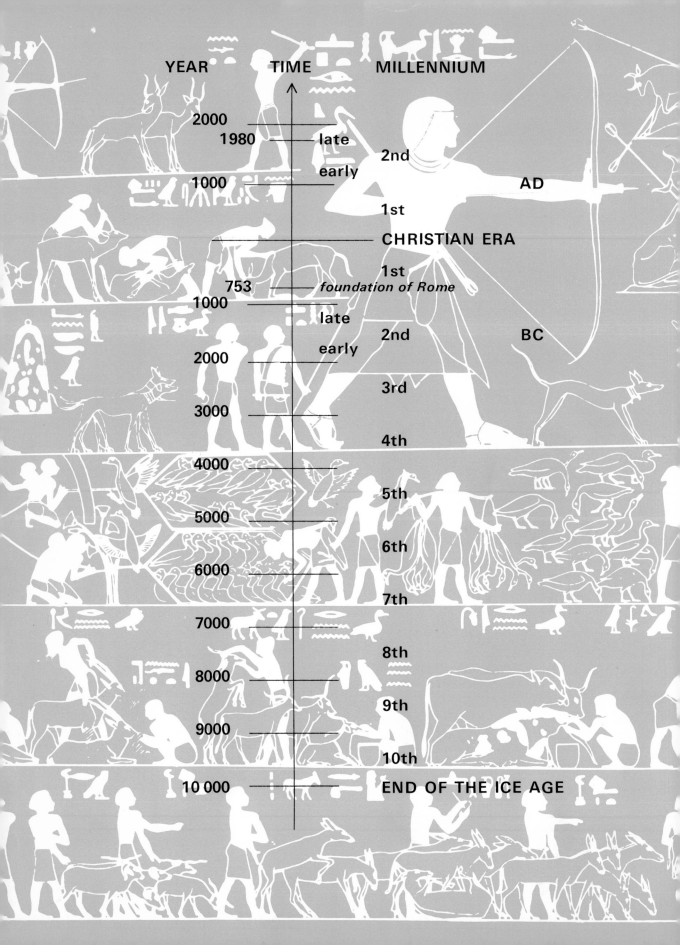

YEAR	TIME	MILLENNIUM
2000		
1980	late	
		2nd
	early	
1000		AD
		1st
		CHRISTIAN ERA
		1st
753	foundation of Rome	
1000		
	late	2nd
	early	BC
2000		
		3rd
3000		
		4th
4000		
		5th
5000		
		6th
6000		
		7th
7000		
		8th
8000		
		9th
9000		
		10th
10 000		END OF THE ICE AGE